QUICK TIDINGS OF HONGKONG

by the same author

Invitation to an Eastern Feast
Personal and Oriental
The Road
Basutoland
City of Broken Promises
Rizal, Philippine Nationalist and Martyr
Myself a Mandarin
Western Pacific Islands
China, India and the Ruins of Washington
Islands of the South
Numerology
A Mountain of Light
A Macao Narrative
Whampoa Ships on the Shore
China Races
The Commerce in Rubber
Macao and the British
(*previously published as* Prelude to Hongkong)

QUICK TIDINGS OF HONGKONG

AUSTIN COATES

Illustrated by
John Warner

HONG KONG
OXFORD UNIVERSITY PRESS
OXFORD NEW YORK
1990

Oxford University Press

Oxford New York Toronto
Petaling Jaya Singapore Hong Kong Tokyo
Delhi Bombay Calcutta Madras Karachi
Nairobi Dar es Salaam Cape Town
Melbourne Auckland

and associated companies in
Berlin Ibadan

First published 1990
Published in the United States
by Oxford University Press, Inc., New York

ISBN 0 19 584024 0

British Library Cataloguing in Publication Data available

Library of Congress Cataloging-in-Publication Data available

Printed in Hong Kong by Nordica Printing Company
Published by Oxford University Press, Warwick House, Hong Kong

Preface

FOR more than a hundred years Hongkong's various forms of quick communication — telegraph, telephone, wireless — were provided and run by various different international and local concerns. It had never really been satisfactory. Services which are fundamentally interdependent were conducted on a basis of rivalry, and frequently of mistrust. But it was in this way that things had developed, owing to the past, to events only dimly recalled, and to people whose very names had been forgotten.

In 1984 a change occurred. Cable and Wireless, established in Hongkong since 1938, became responsible for the telephone system. Thereby, for the first time ever, the whole of Hongkong's telecommunications came under unified control.

This change was of so radical a nature that it was felt by those concerned to be the end of one era and the beginning of another. From this arose the sensation that it was an appropriate moment to tell the story of the old era, while memories of a part of it were still fresh, and while written records still existed. The outcome is this historical study of Hongkong's telecommunications — quick tidings, which is a translation (one of several alternatives) of the two Chinese characters which convey this word.

I should explain at the outset that what follows is strictly non-technical. The aim is to tell the story of what happened in the past, and in places it is quite a strange story. The succession of different companies, and the relations between some of them, are at times confusing. On pages *xiii–xiv* there is a chronology intended to make these aspects plainer.

As principal documentary source for the period 1865–1933 I used the Colonial Office papers (they include Foreign Office material) relating to Hongkong, series CO 129, of which a complete photostat set, admirably indexed, is held at the Public Records Office, Hongkong. As my other principal source I used the board meeting minutes, often with copious and interesting correspondence as enclosures, of the Hongkong Telephone Company. These date from 1925, and are complete.

Printed and published sources used included (in the Government Secretariat library) the pre-1941 Administration Reports and Legislative Council proceedings, the post-1945 Hongkong Annual Reports, the Postmaster-General's annual reports, the proceedings and findings of various commissions and departmental inquiries; and the colony's newspapers, which are in the Public Records Office.

Two books were consulted: *Girdle Round the Earth*, the story of Cable and Wireless and its predecessors, by Hugh Barty-King (Heinemann, London, 1979), a masterly handling of a very large subject; and *The Far Eastern Telegraphs*, by Jorma Ahvenainen (Finnish Academy of Sciences, Helsinki, 1981), on which I drew for the early story of the land telegraph in Russia and Siberia.

Another source was an unpublished history of the Hongkong Telephone Company written in 1965 by Stephen Grove, who was for many years the company secretary and joint managing director. His historical knowledge of old Hongkong was insufficient to enable him to unravel the mystery of the first two telephone companies (about which there are no records), but his observations and comments on the third one seemed to me to be extremely sound, and I have quoted him in various places.

I wish to express my thanks to Professor C. Mary Turnbull, at that time head of the Department of History in the University of Hongkong, for drawing my attention to Jorma Ahvenainen's book, and for very kindly lending me her own copy of it; to the Revd Carl T. Smith, who has for years been gathering documentary material concerning people and places in Hongkong's past, and who helped me to shed a clearer light on certain people who come into this book; and to past and present members of the management and staff of Cable and Wireless and the Hongkong Telephone Company who provided me with information which I would not otherwise have been able to obtain.

The sponsors will, I know, wish to join John Warner, the illustrator, and myself in thanking for their goodwill all those who allowed pictures, engravings and photographs from their private collections to appear in these pages.

Contents

Part IV Hongkong Speaking

Plates

Maps

Companies and Dates

Relative to Hongkong

1871 Telegraph service by Great Northern Telegraph China and Japan Extension Co. (subsidiary of Great Northern Telegraph Co., Copenhagen)

and (separate but with a shared office)

China Submarine Telegraph Co., London (a component of what became the Eastern Telegraph Group of Companies)

1873 Eastern Extension Australasia and China Telegraph Co., London (China Submarine Telegraph Co. merged with two related London companies — Eastern Telegraph Group)

1876 Graham Bell's invention of the telephone

1883 Telephone service by Great Northern Telegraph Co., Copenhagen (amalgamated for Hongkong telephone purposes with Oriental Telephone Co., London, founded by chairman of Eastern Telegraph Group); service closed down within the year

1890 Telephone service by China and Japan Telephone Co., London (wholly owned subsidiary of Oriental Telephone Co., London, with same chairman and directors)

1896 Guglielmo Marconi's first patent for wireless telegraphy

1915 Wireless service by Marconi's Wireless Telegraph Co., London, shortly taken over by Royal Navy

1921 Wireless service by Hongkong Post Office

1925 China and Japan Telephone Co., London, replaced by Hongkong Telephone Co., Hongkong

1928 Telegraph service by Imperial and International Communications Ltd., formed by merger of the Eastern Telegraph Group of Companies, London (including Eastern Extension Australasia and China Telegraph Co., which retained its name) with Marconi's Wireless Telegraph Co., London

1934 Imperial and International Communications Ltd. changed its name to Cable and Wireless Ltd.

1938 Wireless service transferred from Hongkong Post Office to Cable and Wireless Ltd. Eastern Extension Australasia and China Telegraph Co. changed its name to Cable and Wireless Ltd.

1947 Cable and Wireless Ltd. nationalized, with British Government as sole shareholder

1981 Cable and Wireless Ltd. denationalized; a local company, Cable and Wireless (Hongkong) Ltd., formed

1984 Hongkong Telephone Co. became an affiliate of Cable and Wireless Ltd., London

1988 Hongkong Telecommunications Ltd. formed by merger of Hongkong Telephone Co. and Cable and Wireless (Hongkong) Ltd.

PART 1
A Gradual Approach

So gradual has been the approach of the telegraph to Hongkong that people almost think of it as quite in the ordinary course of events, and look with astonishing complacency upon the accomplishment of a scheme which, ten years ago, it would have been considered almost madness to hint at.

Hongkong Daily Press
6 June 1871

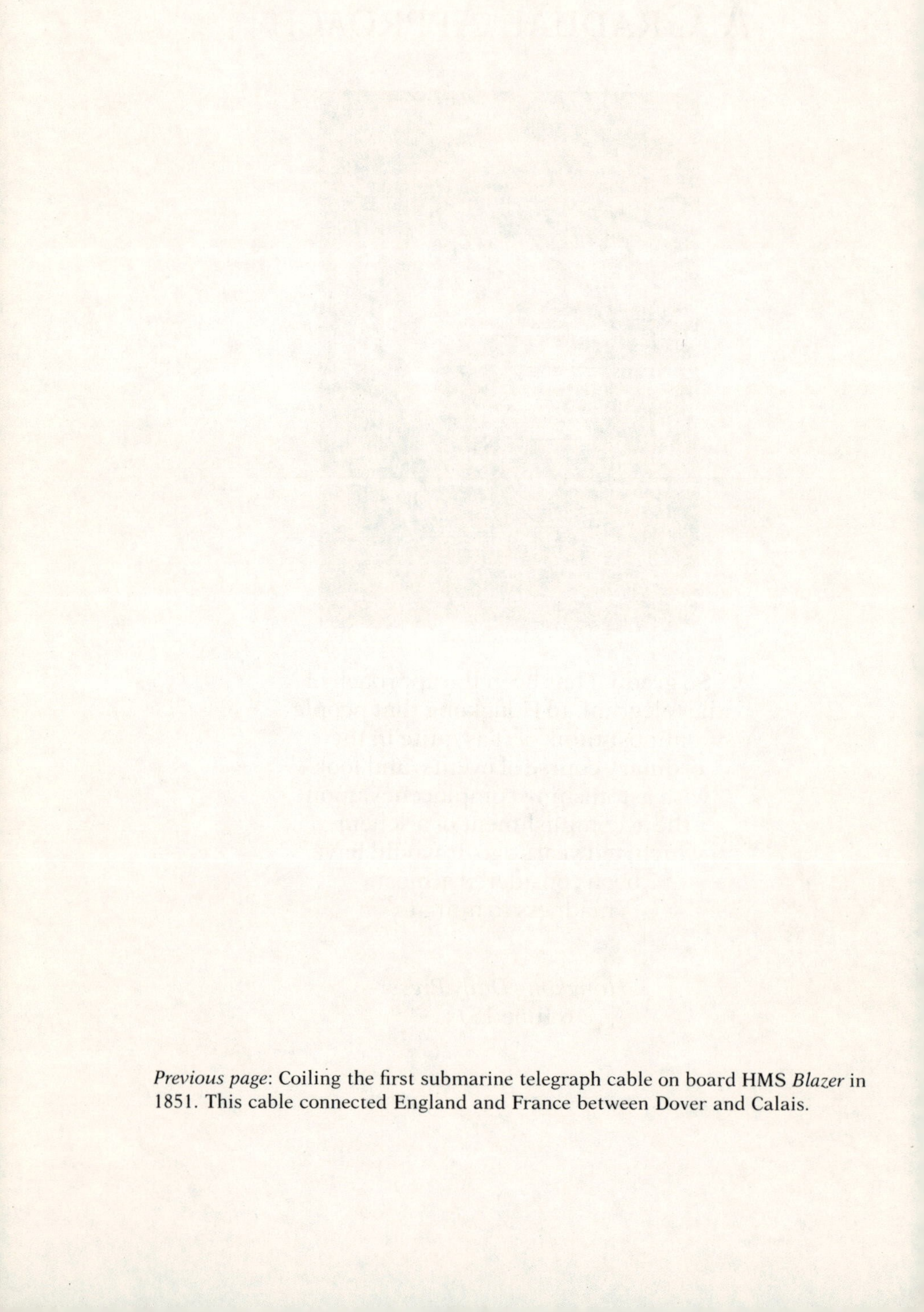

Previous page: Coiling the first submarine telegraph cable on board HMS *Blazer* in 1851. This cable connected England and France between Dover and Calais.

Chapter 1

In India and into Asiatic Russia

EARLY in December 1865, the Acting Governor of Hongkong — the Governor was away — received a letter from an American in Shanghai.

The American sought permission to land a submarine cable on Hongkong Island, and a concession to operate a cable telegraph service between Shanghai and Hongkong. Since the distance between these places was greater than any cable could be expected to link, there would be two landings on the route, at the Treaty Ports of Amoy and Foochow, with both of which a fair amount of telegraph traffic could be expected.

Decision on this letter required reference to the Colonial Office in London. Considering the matter to be of sufficient importance, the Acting Governor decided that it warranted the expense of a telegram.

The telegram left Hongkong by the P. & O. fortnightly mail steamer, destination Suez. After a 48-hour call at Singapore, and a 24-hour call at Penang, the steamer headed for Colombo, its next call. Passing the southern tip of Ceylon, with nine hours to go before reaching Colombo, the steamer narrowed to the ancient port of Galle, where a small steamboat came out and took the telegraphic material, among which was the Acting Governor's telegram.

At the Galle post and telegraph office it reached its real element, which was land, not sea. Tapped out in the Morse code, along with all the other telegrams, it moved at incredible speed, covering the greatest distance in the shortest time known in the history of mankind, and reached London in six days.

The service had been in operation for ten months. It reduced the time taken to transmit a message between Hongkong and London from forty days to fourteen days.

The Colonial Office replied that it would be inappropriate for the Governor of Hongkong to grant a concession 'for laying a cable in the China Seas'. (It was assumed, for want of explicit information to the contrary, that this was a matter for the

Emperor of China.) Arrangements should be limited to a grant of a landing place on the south side of the island, away from the harbour, and in return perhaps government messages could have priority. In short, the British Government, being helpless in the matter, did not mind.

It was helpless because anything telegraphic east of Galle foundered on the cable distances.

The telegraph was a land instrument. A telegram from London to the East, with the exception of various submarine cable sections, as between England and France under the Channel, and at Constantinople under the Bosphorus (sometimes, otherwise by ferry), moved at the rate of seven words per minute along thousands of miles of poles and wires, with millions of birds sitting on them at certain seasons.

For hundreds of miles in some parts of the route, the line of telegraph poles was the only noticeable feature in landscapes of utter emptiness, a meaningless feature to a man on a camel, whose sole reaction was to wonder why soldiers would be sent if he damaged it. For he was no wiser than the birds about this strange, solitary, thin black line, heading straight — always straight — from one side of the horizon to the other.

The London telegram to the East moved along this line by agreement with the Imperial French (Napoleon III) telegraph service, the Imperial Austrian service, the service of the Ottoman Sultan, and the Government of Bombay Posts and Telegraphs (more homely somehow), until it reached Galle. There it was franked by the Ceylon colonial postal authority, becoming in effect a letter, and was sent on by ship without further charge.

India had its own telegraph system long before there was any telegraphic contact with Europe. It was imposed, in the teeth of near-hysterical Hindu opposition, by fiat of the Governor-General, Lord Dalhousie, in 1854. It was one of the largest telegraph systems in the world.

Apart from supplying equipment, Britain had little to do with it. Neither Britain nor any other European nation could reach India telegraphically. When at last this became possible, it was found (where it was not already known) that of the two potentates controlling the region between Europe and India — the Ottoman Sultan and the Shah of Persia — dealings were feasible with one or the other, but not both.

The Sultan, whose dominion extended well into Eastern Europe, was the more important of the two, and he was interested in innovations, which was more than could be said of the Shah. Arrangements were accordingly made with the Sultan, and the Shah's dominions were bypassed — by a submarine cable.

The cable ran from Gwadar on the coast of Baluchistan, in British India, to the head of the Persian Gulf, where the link was made with the telegraph service of the Ottoman Empire. There were two cable landings on this section, much as was being suggested for the Shanghai cable. Anything longer than this was a perilous financial enterprise. The Atlantic cable linking Europe and America was opened for service in 1858 with thunderous acclaim, but broke down after five weeks and had not been replaced. Various attempts to lay cables across even such relatively short distances as the Mediterranean had failed. Enormous sums of money had been invested and lost on cables.

More practical from Hongkong's point of view, and Shanghai's, was the overland telegraph. The originating genius of this was the Danish banker T. F. Tietgen, director of Privatbanken, Copenhagen, who in the early years of the telegraph conceived the idea of linking Europe and America, and by the same token Asia, on a northern circuit which would bypass the European Great Powers — Prussia, Austria, and France. Their propensity to engage in wars with each other was a standing liability to an international telegraph service which passed through any of their territories.

From Copenhagen, the eastward advance of this concept passed by submarine cable to Sweden, thence by land, northward to the northernmost point of the Gulf of Bothnia, where it entered Finland, southward and again eastward through Finland, until it linked with the Imperial Russian telegraph system and the Russian capital, St Petersburg.

The link with St Petersburg was made at a time when Russia's opening towards the East, into Asiatic Russia, received a dramatic new impulse. In 1860, as an outcome of the Second Chinese War — actually fought against China by Britain and France, but in which the tacit moral support of Russia and the United States was crucial — China was obliged to cede to Russia the Maritime Province, giving Russia an extended North Pacific coastline on which she proceeded to develop the naval port of Vladivostok. Telegraphic communication between St Petersburg and Vladivostok being considered desirable, the advance of the telegraph eastward across Asiatic Russia proceeded with more verve. With it came the extending Danish system and Danish technical advice.

There was another influence working for a northern telegraph circuit. Perry Collins, an American who as a lawyer and banker had made a fortune in the Californian gold rush, had formed the idea of linking Europe and America across the Bering Strait, where the distances between points of land were within the viable limit of cables. The American telegraph was advancing westward across the United States; the Russian telegraph was slowly extending eastward. Across the Bering Strait they would be connected. In 1856, having established an entrée in St Petersburg, Collins had himself appointed United States Commercial Agent in the Amur river region of Siberia.

The failure of the Atlantic cable in 1858 gave assurance that he was thinking on the right lines. Russia's acquisition of the Maritime Province, and the consequent hastening of the telegraph's eastward advance, brought the matter closer to realization. In 1863 Russia, satisfied that Collins had the support of the authorities in British Columbia, a territory integral to the scheme, formally approved the Bering Strait proposition. It looked as if it was to be a Danish-Russian-American undertaking.

It would take a few years before the link was complete and making money. Of immediate concern was the vast length of telegraph line through almost uninhabited Asiatic Russia, which would be profitless until the Bering Strait link was made, and in its own right would be profitless even after that. Another source of revenue was needed, so in 1864 Perry Collins proposed running a telegraph line to China, which was certain to be a revenue-earner.

The line would have its juncture with the main Russian line at Kyakhta on the northern border of Mongolia, south of Lake Baikal. Kyakhta was the head of one of the most important caravan routes in Central Asia, a route which led through Outer and Inner Mongolia directly to Peking. From there Collins envisaged a telegraph system which would link Nanking, Shanghai, Amoy, Canton, and Hongkong.

This arrangement met with Danish approval, and among the reasons for this was a special one. The Danish banker Tietgen had originally envisaged Britain as one of the Great Powers to be avoided in his northern system. He had changed his mind. The extent of Britain's power in the world — power greater than that of any other nation — was such that a large international organization, such as a telegraph system, stood a better chance of succeeding if there was a British interest in it. If the overland telegraph were brought to Hongkong, and provided that the correct moves were made in Copenhagen and London beforehand, there would be a very definite British interest, which would be advantageous. Hongkong was the most remote of all Britain's outposts in Asia, the uttermost end of the line.

Disconcertingly, the Chinese authorities, when approached on the proposition, refused to countenance it, politely but flatly explaining that the population at large were likely to be antagonistic to the telegraph, and that in consequence it could not be admitted into China.

Danish instinct in matters Chinese, derived from a hundred years of trading with China, said that a negative response from a Chinese official source did not always mean quite what it said. One could bide one's time, then try the same thing again in a different way. The telegraph would in any case reach the Pacific. From that point a chain of cables down the Korea and China coast would achieve the same revenue-bearing effect. Since this would not be on the soil of China, other than at unobtrusive landing places, perhaps the Chinese official attitude would be less adamant. The suggestion of it might even make the Chinese change their minds about their opposition to the telegraph.

This idea, before anything had been done about it formally, passed in the strange way ideas do — probably from something Collins said in a letter — to another wealthy American, James Milliken, who was playing with the idea of investing in cable telegraphy. As he saw it, the best place to start would be on the section most certain to bring an immediate revenue, which was the section between Shanghai and Hongkong.

This was what lay behind the Shanghai letter to the Acting Governor of Hongkong. It was to be the first step in an attempted cable extension of the overland telegraph from Copenhagen, and neither the Acting Governor nor the Colonial Office — nor the Treasury, who were also consulted — realized this. The geographical dimensions were too large, the ideas too bold and unexpected, and the scene of action lay in lands of total mystery.

The Governor's reply to the letter reached Shanghai late in March 1866 — a reply in four months, which considering that the matter had been dealt with by a colonial government and two ministries was very quick. By then, however, a new feature was discernible beyond the horizon.

For some months there had been reports — all of them received on the China coast two or three weeks late — that another attempt was to be made to lay a cable across the Atlantic. Early indications of the event were confused. The Prince of Wales had seen a ship off; a cable had been sabotaged; a company had collapsed, or nearly collapsed. It did not sound inspiring, yet the report persisted; by the spring of 1866, people listened.

At the head of the Russian telegraph, news was received more quickly. A cable across the Atlantic, from Ireland to Newfoundland, would mean telegraphic communication between London and New York. If the cable held, it would be the end of the Bering Strait scheme, which had the same aim: London to New York. Who would use it, if there was an Atlantic cable?

In August the news that the *Great Eastern* had laid an Atlantic cable came to Siberia. After the usual time-lag of two to three weeks the news reached the China coast. There, at all the usual places — the Race Club at Shanghai, the Hongkong Club, and so on — there was always one man who knew exactly how many days the first Atlantic cable had been in operation. When it was realized that the new cable had held successfully for one day longer than the former one, there was complete and totally irrational conviction that the new one was there to stay. And for once, an irrational conviction proved to be right.

Having survived the man who said 'I told you it would', heads turned to wonder what cable development would come next — east of Galle, that is to say.

Chapter 2

The Impediment of China

PERRY COLLINS' inevitable withdrawal from the Bering Strait venture, and from the attempt to bring the telegraph into China, left the field free for the Danes. It was idle to pretend that Russian powers of persuasion would have any effect on imperial Peking unless to inspire fear, whereas the Danes, representative of a small nation long known by the Ch'ing court to have no hostile intent, might succeed if they persevered. This they had every intention of doing.

A more careful assessment of Chinese unwillingness to allow the telegraph into China tended to the view that the real obstacle was less the possible antagonism of the populace at large than the consideration that the line would start at Kyakhta.

Russia had already acquired the Maritime Province. As Peking would see it, the next region on which Russia had designs must certainly be Outer Mongolia. The telegraph, despite the Danish involvement in it, was a Russian telegraph. For its poles and wires to run across Mongolian territory would be the harbinger of claims and demands for rights, which could themselves be the precursors of territorial claims. Under no circumstances, the Danes concluded, would China allow a line to enter from Kyakhta; and as the succeeding forty years were to show, this conclusion was accurate. Despite its being by far the shortest and most convenient telegraph route, every attempt to open it would be balked.

This Danish conclusion led to further thought regarding the alternative: a system of submarine cables down the China coast. There was the added commercial compulsion that with the Bering Strait venture in abeyance, the eastern telegraph, across the vast uninhabited wilds of Siberia, had little prospect of profitability unless it extended to Shanghai.

As the success of the Atlantic cable demonstrated, a great advance had been made in the technique of cable-making and cable-laying. With stronger cables linking longer distances, the

China coast cable idea, which had at first been hardly more than a pipe dream, had become technically — though not by any means politically — a realizable proposition. Above all, it would geographically dispose of the most obdurate sector politically, which was Korea, at that time a completely enclosed country, the Hermit Kingdom. In the new circumstances, a submarine cable could be laid directly from Vladivostok to the Treaty Port of Nagasaki, in Japan, from which another cable could be laid directly to Shanghai.

In Japan itself there might be the likelihood of an overland telegraph from Nagasaki to Yedo, the capital, soon to be renamed Tokyo. By laying another cable from Nagasaki to the mouth of the Peiho, in China, and possibly up the river to Tientsin, even if China still would not allow a land telegraph, this would bring the system to within eighty miles of Peking, near enough for it to be used by the diplomats. The Peking diplomatic traffic could be expected, at least at first, to be almost as lucrative as the Shanghai commercial traffic.

The actual cable-laying for this system, and the initial obtaining of rights, would have to start at Hongkong, as in the case of the earlier American proposal, about which nothing more had yet been heard.

This Danish grand design, to link Europe with Japan and China through Russia, would require a concession from the Russian Government. Here the Great Northern Telegraph Company, the Danish organization concerned, had the advantage of enjoying the full support of the Danish Government, including Danish diplomatic support in St Petersburg.

Still, as only a casual glance at the scheme showed, if it was successfully brought into operation, it would be a highly attractive commercial undertaking, providing quick contact with the two most important and most aloof capitals of the Orient. Even the Russians, not a people particularly noted for commercial acumen, would realize this, as also that they themselves were in a position to hold the scheme to ransom unless they received an attractive share in it. The inescapable position was that, seeing their own country as poor, with wealth to the West and wealth to the East, and themselves in the middle, the Russians would seek to lay their hands on far more of the profit than their contribution to it warranted.

Which they did. While the overland telegraph painfully advanced to the Pacific, the Danes were engaged in negotiations of exceptional difficulty, only one degree less difficult than dealing with China. They succeeded — but not before October 1869 — in reaching agreement at 60 per cent to the Russian Government, 40 per cent to Great Northern, on all traffic passing through Russia, and were granted the required concession. Although the Russian final figure was still unwarrantably high, this ranks as an outstanding piece of commercial diplomacy.

Within days of the signing in St Petersburg, Great Northern announced in Copenhagen the formation of a subsidiary, the Great Northern Telegraph China and Japan Extension Company. A few weeks later, through their bankers C. L. Hambro & Son, shares were offered in London.

The company's share capital was £600,000 in 60,000 shares at £10 each, of which 45,000 shares were offered in London, the remainder having already been subscribed in Copenhagen. The prospectus, after explaining the aim of connecting China and Japan with Europe through Russia, concluded with the words: 'The peculiar and commanding advantage of this route is its shortness.'

This feature — the shortness of the route — was a prime card for Great Northern in relation to another company, formed or about to be formed, which they knew about. It was a British company which aimed to link Europe with China and Japan by the sea route.

From the moment when it had finally dawned on business men and investors in Britain that the Atlantic cable was there to stay, that a great technical advance had been made, and that the submarine cable was now a reliable conduit over long distances, activity in the cable business, and keenness to invest in it, had generated a state of affairs which the Foreign Office described as 'cable mania'.

On the investment side, it had been facilitated by the British Government's decision to acquire by purchase the several dozen private telegraph companies on which the country was entirely dependent for this service, bringing them all under the ownership and management of the Post Office. Government purchase extended to the 'domestic' cable companies, linking Britain with Ireland and the Channel Islands, and to the Dover-to-Calais cable, the world's first submarine telegraph (1851). As a result, a mass of owners of 'domestic' cable and telegraph stock found themselves with handsome sums of cash, most of which went, rather logically, straight into the newly forming overseas cable companies.

Raising money for cable ventures had become progressively

more difficult as one cable after another failed. Shortage of money nearly wrecked the attempt to lay the Atlantic cable in 1866. Its patent strength and durability, the efficiency of its operation, changed investors' attitudes. The unexpected windfall from Her Majesty's Government came as the benison. At last there was money. The world's communications proceeded to improve at extraordinary speed, not solely in the direction of the East, but on all the major routes of human enterprise.

A British company formed in 1868 successfully laid a cable between Malta and Alexandria. They at the same time purchased two existing shorter-distance cables, between Malta and Sicily, and between Sicily and Italy, thereby creating a link which dramatically speeded up telegraphic communication between Britain and Egypt. This improvement had no effect on Hongkong, of course, where telegrams continued to pass to London as they did when first encountered.

Next, in January 1869, the British-Indian Submarine Telegraph Company was formed, to lay a submarine cable from Suez down the Red Sea to Aden, and under the Arabian Sea to Bombay. This would reduce the time it took to send a message from Hongkong to London from fourteen days to eleven days — provided that one sent one's telegram on the day the mail steamer left.

The chairman of British-Indian Submarine, and its driving influence, was John Pender, a Scottish business man who had early made a fortune in Manchester cotton, and married an heiress, Emma Denison of Nottingham, a woman no less remarkable than himself. He had been fairly prominent as a director and shareholder in some of the earlier cable companies, notably the Irish one, but first came to real prominence in the cable world in 1866. When it looked as if the Atlantic cable venture was going to fall through, really from lack of confidence more than anything else, at the last moment he guaranteed it to the tune of £250,000 of his own and his wife's money. It saved the situation, and when the venture succeeded, cables thereupon became John Pender's life. In the years ahead he was to become known with unstinted respect in all parts of the world as the Cable King, which indeed he was.

Directionally speaking, his first thoughts lay towards the East, and he thought imperially, a manner of thinking with which the Prince of Wales, who was on occasion entertained by the Penders at their London home in Arlington Street, would have agreed. The imperial concept was still comparatively new in Britain, and it was entirely genuine. There was none of the hollow ring about it which it was to acquire in the next century. Having arranged for his cable to Bombay, Pender's follow-up thought was of the need for Britain to have an imperial telegraph of her own, independent entirely of any other country, a cable telegraph.

Five months after founding British-Indian Submarine, he formed another company to lay and operate a cable from Falmouth to Malta, with stations at Lisbon (for a South American cable in due course) and Gibraltar. By coming to a financial and operating arrangement with the British company running the section between Malta and Alexandria, and this coinciding with the completion of Ferdinand de Lesseps' monumental work on the Suez Canal, Britain and India communicated directly on an entirely British imperial line.

This, parenthetically, reduced Hongkong's London message time from eleven days to ten days.

Quite clearly, with John Pender in action on the grand scale, the imperial telegraph line would extend further into the East, the first objective being the Straits Settlements, and in particular Singapore. From there it would be possible to bring the cable telegraph to China, and by arrangement with the Dutch authorities in Batavia (now Jakarta) and The Hague, lay a cable, or a series of cables, along the islands of the Netherlands Indies and achieve what it had always been assumed would never happen, linking London telegraphically with the Australian colonies, which already had the beginnings of their own internal telegraph.

China, however, was foremost in his thinking at this time; and China, to a man of his commercial sense, meant Shanghai. Although Hongkong was important as an outpost of empire, and would be the first point to be reached, it was only a small place in terms of commerce. Shanghai was where the business was, and where the traffic would be.

He decided, with the approval of his fellow-directors, to form a British-Indian Extension company to handle the cable eastward from Galle — in due course it went from Madras, which was a shorter route — to Penang, thence down the Strait of Malacca to Singapore. For the extension beyond that point he resolved on yet another company, the fourth major concern to be founded by John Pender in that one year 1869, the China Submarine Telegraph Company.

This manner of proceeding reflected Pender's style. The

amount of capital required for a cable company was large by any standards. So was the organization which was needed to lay, maintain, and operate the cable. Each undertaking was a major land-and-sea administrative operation. John Pender's tactic was to proceed company by company, each addressed to a clearly defined geographic sector, each launched on the basis of a reasonably attainable amount of capital, each with a zone which would be operationally manageable. In making the division between his two new companies at Singapore, however, there was a further consideration, which was the element of risk.

The British-Indian Extension company, formed to link India to Singapore by cable, was a sound scheme presenting no impediments and only marginal risk, provided all went well technically. The entire undertaking lay within the British imperial frame.

Beyond Singapore was a hazard, with which shareholders in the 'safe' extension must not be concerned. Those who chose to be concerned must go into it with their eyes open, in a different company.

It was a hazard because unless the cable reached Shanghai, the extension was unlikely to make any profit. Unlike Hongkong, which was a British colony, Shanghai was a Treaty Port, meaning that while foreigners whose governments had concluded treaties with China were exempt from Chinese legal process and had their own municipal administration, Shanghai itself was under Chinese sovereignty.

China's attitude to the telegraph — the attitude of the Ch'ing court at Peking — was in fact even more difficult than the Danes had conceived it to be. It was in very truth adamant in its opposition to it. Nor was the postulated antagonism of the populace to the telegraph a factor to be dismissed lightly. When the Chinese authorities referred to the populace, they meant the gentry, the one or two per cent who were literate and educated, small in numbers but large in influence, bastions of conservatism, foundations upon which the Throne rested.

The Throne was at this time probably even more wary of the unknown than the gentry were; but the traditional line of thought was the same. Once it was understood what the telegraph was — and the Throne did vaguely understand what it was — it would arouse disquiet, because it seemed to be contrary to nature. This was not the thinking of an outlandish tribe; it could have been dismissed if it had been. It was the reasoned thinking of an old civilization, and for that reason all the more complex.

Not that it would have interested anyone in China to know it, but in India it was from precisely the same segment of the population, and for precisely the same reason, that the violent Hindu opposition to the telegraph had arisen. Lord Dalhousie ignored it, calling out the troops when necessary, and in a surprisingly short time the Hindu élite discovered with satisfaction what a wholesome convenience the telegraph was.

The difference which confronted John Pender and his board of directors in the year 1869 was that China did not have a Dalhousie.

Chapter 3

A Delicate China Venture

BY the standards of the time, and considering that very few Westerners had ever been to the Far East, John Pender was well informed about what to expect beyond Singapore. Among the directors of his projected China Submarine Telegraph Company were Horatio Lay, former head of the Chinese Imperial Maritime Customs (run by foreigners on China's behalf, and the principal source of China's revenue), and the millionaire Reuben David Sassoon, whose family's business ramifications were continental in scale, particularly strong between Bombay and Shanghai.

Other directors bore names with a more familiar ring to them on the London scene: Lord William Montagu Hay, and William Massey, who was a Privy Councillor, each of them chairman of one of the companies making the imperial cable link; then two prominent City men and the former Director-General of Indian Telegraphs. Where special information was concerned, however, it was the less familiar names, Lay and Sassoon, at the foot of the list, who mattered.

Pender's cable activities brought him frequently in touch with Foreign Office officials, another source of information, and more recently with the Colonial Office, then at its zenith, one of the five senior ministries. He himself was for much of his life a Member of Parliament, which gave him a standing additional to that of a prosperous business man. Thinking imperially — globally, for that matter — gave a certain tone to proceedings at the Pender home in Arlington Street. Sir Robert Herbert, Under-Secretary of State for the Colonies and a very senior civil servant, was sometimes to be seen there — occasions which he clearly enjoyed. Emma Pender, in the nicest possible way, referred to him as 'Herbert of the Colonial Office', rather as if he were the assistant butler.

Yet when John Pender intimated to the Foreign Office that British diplomatic support in Peking for the landing of cables in China would be in Britain's national interest, he was courteously disabused of any hopes of such assistance. It was explained to him that he would have to make all the overtures himself, through his own agents. The truth was that the Foreign Office would have been only too thankful if the telegraph came to China — British investment in China was greater than that of any other nation — but Britain had had too many embroilments with China already. To lend official support to the commercial introduction of an innovation which China did not like, and was certain to resist, would simply be asking for more trouble.

Pender's colleagues on the embryo board of China Submarine seem to have been confident, however, that somehow the cable would reach land at Shanghai. The presence of Horatio Lay among them was probably decisive in this respect. A controversial character to Britons in China, he was a resourceful person with wide experience of China and a way of getting things done in that country, not always by methods considered orthodox by others. A plan of how to proceed must have existed, though what it was can only be surmised. Where raising capital was concerned, there was one essential. Facts, and plenty of them, about the Far East must be presented to the public, leaving no doubt of the company's assured prospects in that region.

The share capital of the China Submarine Telegraph Company was to be £525,000, in 52,500 shares of £10 each (nothing to do with cables was ever a poor man's investment), of which a total of 10,000 shares were to be taken up by the four cable companies creating the imperial telegraph line, three of them Pender companies. It was felt that this would be of assurance to the public (which it was) in what was by cable standards a distinctly delicate business.

Then came the facts: 486 English and foreign firms were established in Hongkong and the Treaty Ports; there were 486 Chinese firms dealing with foreigners in Hongkong alone (if they had stated it as 'firms and little shops' it would have been nearer the truth); six-and-a-half million tons of foreign shipping entered and cleared Chinese ports the previous year, in 14,075 ships, and the total value of imports and exports based on the Customs returns (conveniently provided by Horatio Lay) was £68,000,000. These (undeniably impressive) figures, the prospectus explained, did not include the very substantial local trade (meaning trade in Asia, which was in fact where the greater part of the British investment lay, and had nothing to do with the British national economy). Thus, in addition to telegraph traffic with Europe and America, messages between China and India, the Straits, and the Eastern Archipelago would constitute a major part of the revenue, producing 'a very extensive Telegraphic business ensuring large dividends to shareholders'.

The route was given in the prospectus the whole way from London. From Singapore the route would be by sea, it was explained, probably to Saigon, thence to Hongkong, and from there, 'it is hoped', to Amoy, Foochow, Shanghai, Tientsin, and Nagasaki, by land from Tientsin to Peking, and by land from Nagasaki to Yedo.

All of this was to be in comparatively small print. The large print would come at the end, with no 'it is hoped' attached to it: 'FROM SINGAPORE TO SHANGHAI, AND INTENDED EXTENSIONS TO PEKING AND YEDO.'

The reader will appreciate what is meant by this being a delicate business. Their chances of reaching Shanghai were marginal, of reaching Yedo minimal, and of reaching Peking nil.

Moreover — and whether all the directors knew this is doubtful — the facts presented were misleading. By generalizing the facts and figures, and by several times using that misleading style 'Hongkong and the Treaty Ports', a calculated impression had been created that Hongkong was an important place commercially, which it was not. To create such an impression, however, would be of assurance to the type of investor who reads the small print, and who would reflect with satisfaction that Hongkong was the one place in the Far East where the British flag flew as of right, and not by courtesy — the type of investor who would feel safer with his money after finding that a Hongkong postage stamp bore the Queen's likeness.

In fact, the one really important place in the Far East at that time was Shanghai, which it was essential to reach if the cable company shareholders were to receive dividends. Next in importance after Shanghai was Foochow, the principal commercial centre and port for China tea, which was still at that date the largest and most valuable single item in the whole of Asia's exports to the West. Yet in the prospectus, both Shanghai and Foochow were in the 'it is hoped' segment.

The prospectus, masterly in its way, bore the distinct imprint

New York celebrates the Atlantic telegraph, over which Queen Victoria and President Buchanan exchanged messages on 13 August 1858. Transmission weakened and expired after five weeks.

Cartoonists in England had a field-day. Father Neptune, *right*, uses the Atlantic telegraph as a clothes line.

The P. & O. passenger and mail steamer *Mooltan*, one of the vessels which bore Hongkong telegrams to, *below*, the town of Galle in southern Ceylon, where they joined the telegraph system.

of having received Horatio Lay's guidance. One was left to hope it would not do China Submarine any harm.

Preparations for the launching of this company, and its prospectus, were already in hand when the wealthy American, James Milliken, reappeared on the scene, in London this time. Having realized from his earlier experience that dealing with the Governor of Hongkong was a waste of time, because that functionary, in any matter concerned with cables, merely did as he was directed by the Foreign Office or the Colonial Office, Milliken had come to London to address himself to these ministries in person. For this purpose he had taken residence in George Street, Hanover Square, a piece of finesse which showed him to be one of those Americans who understand London exactly, for it was a perfect choice.

His immediate aim on this occasion was to make application for running a cable between Shanghai and Hongkong and, 'if all goes well', extending to Singapore and thence to Galle, to link in due course with the 'imperial' cable, which was not yet through to Bombay at that date.

Then, in the last days of November 1869, the Madrid Gazette announced that the Spanish Government, through the Governor-General of the Spanish Philippines, had granted a forty-year concession to two Americans to lay cables and operate a telegraph service linking the Philippine Islands with Singapore and Hongkong. This reached the Colonial Office through the columns of *The Times* in London.

Four days after it did so, the Danish Minister in London approached the Foreign Office, asking them to give a fair wind to a Danish cable company which would shortly be seeking permission to land a cable in Hongkong connecting with Shanghai. Cable mania indeed it was; and so inconveniently close to the Christmas holiday too.

The first to be tackled was the Madrid announcement. This, it appeared, was a monopoly concession, and it was contrary to British official policy to allow any monopoly concern to land cables or operate a telegraph service on British territory.

From the telegraph's beginnings, around 1846, free enterprise had been the British rule, unlike in the rest of Europe (and India), where it was perceived that the telegraph was an instrument of government and must be owned and run by the State. The confusion caused by free-enterprise telegraphy in Britain had resulted in the imposition of a monopoly — that of the Post Office — but with the near-total illogicality with which British public affairs are generally conducted, free enterprise remained the rule.

There was a further consideration in the Madrid case. Within the foreseeable future, British cables would be landed in Singapore, and perhaps Hongkong. While appreciating Spain's desire for cable communication with the Philippine Islands, this must not be achieved by way of a Spanish monopoly of cable communications in two British colonies, which was what it appeared to amount to. The Spanish Ambassador in London was queried on this monopoly point, and since this would mean his referring to Madrid, and Madrid perhaps referring to Manila as well, it would take a pleasantly long time before there was an answer. Whitehall could therefore relax. In the meantime, orders were sent to the Governors of Hongkong and the Straits Settlements not on any account to allow a cable from Manila to be landed. It was the first time either Governor had heard of the scheme. Colonies were not the best-informed of places.

Of the Danish Minister a similar query was made verbally regarding the Danish cable company. The reply received was satisfactory. Though it enjoyed full Danish government support, the Great Northern Telegraph Company was not a monopoly. Foreign Office reaction being entirely favourable once this point had been cleared, the Danish Minister then sounded the ground on the possibility of British Consuls in China assisting with permission for landing cables in that country, and of Her Britannic Majesty's agents (a euphemism for the Royal Navy) affording protection to cables and cable stations.

On this the Danish Minister received a response similar to that given to John Pender. While every reasonable facility would be afforded in Hongkong, which was British territory, none could be given in China, not even to British interests; and the Minister was reminded that under the Treaties it was left that each nation must make its own applications to China. This was exactly so, as the Minister knew. The drawback was that Denmark did not have a Minister in Peking.

As British policy, it was depressingly negative, as Pender had found before. While rigidly excluding monopolies, it allowed that free enterprise was the realizable alternative. It bore no relevance whatever to the situation in China.

The monopoly-hunting part of the policy nearly broke down

when it came to dealing with James Milliken. His was a marvellous scheme. If, as he intended, he linked India and the 'imperial' line with Hongkong and Shanghai — and Canton incidentally — he then proposed to run the cable northward up the China coast from Shanghai, and across the Pacific to North America, thus creating for the first time a world telegraph link.

But, as the bureaucrats noted with anxiety, if he ran his cable south from Hongkong, he would be taking in the opposite direction exactly the same route as the two Pender companies, which had not yet been publicly announced. Here was a formidable competitor, and there was nothing to be done to stop him so long as free enterprise was the policy. If only Pender would launch his companies, the position would be clear to Milliken. But not a sound had come from the Pender stronghold in the City, while to give concessions to Mr Milliken now . . .

And then someone midway up the Foreign Office hierarchy remembered something. James Milliken was American — with that Hanover Square address they had almost forgotten this — and his scheme involved China. Whitehall dealt him the deftest of deft strokes. Under the Treaties, he was informed, the British could not approach Peking except on their own behalf. As an American, he would have to approach the American authorities in Peking to know if China would agree to a cable being run from Hongkong and the Straits Settlements to the coast of China.

James Milliken bowed out, defeated, and very shortly afterwards, to the considerable relief of the Whitehall men who knew the inside situation, the China Submarine Telegraph Company was launched, with its splendid declaration: 'From Singapore to Shanghai, and intended extensions to Peking and Yedo.'

A week after that, Hambro's Bank launched the Danish Great Northern shares (China and Japan through Russia) on the London market, with that telling sentence:

'The peculiar and commanding advantage of this route is its shortness.'

Chapter 4

Anglo-Danish Adjustments

It was a thunderbolt. At least, in the Pender stronghold it was. John Pender and the members of his board, like everyone else in the cable world, knew of the earlier Danish endeavours to make the American link via the Faroes, Iceland, and Greenland. They were aware that a telegraph line had been run through to St Petersburg, and that there had been talk of extending further east. But that was ten years ago. Nothing certain had been heard since then, and not surprisingly. Russia was a sealed mystery to the rest of Europe. What went on there was beyond normal knowledge.

The Whitehall bureaucrats were equally astonished. They knew well enough about the impending Danish application for cable landing in Hongkong, but had utterly no idea that this was part of so vast a design.

At China Submarine, those 'intended extensions' to Peking and Yedo, after the Danish announcement, boomed with a hollow emptiness. In that the Danes had set up this remarkable enterprise as far as they had, and had won the valuable Russian concession, which was an achievement in itself, if anyone was ever to run telegraph lines to Peking and Yedo it would be they, and not China Submarine. For the latter's chairman, a wintry, storm-tossed voyage up the grey North Sea to Copenhagen presented itself to view, to be followed by negotiations in which he would be decidedly at a disadvantage.

The very magnitude of his cable operations, compared with those of the Danish company, in a curious way told against him at this point. It was January 1870. The imperial link between Britain and India would be made in June. After this, two further major links had to be undertaken, beneath the Bay of Bengal to Penang and Singapore, and beneath the South China Sea to Hongkong.

Even with John Pender's speed and sureness of organization, which were superb, it would take at least another year.

By contrast, the Danish company would begin their cable-laying in Hongkong, shipping the cable there via the conveniently opened Suez Canal; and they would reach Shanghai first. This was the point.

It had been assumed that Shanghai would be an assured source of profit, and the assumption was sound enough. The Shanghai commercial traffic would without doubt be profitable, but whether it would be heavy enough to enable two cable companies to operate at a profit was extremely doubtful. Moreover, once Great Northern laid their cables onward from Shanghai to Nagasaki and Vladivostok, Shanghai's telegraph traffic to Europe and America would assuredly be Danish.

In Hongkong the British could be expected to feel more confident availing themselves of a British cable system. There was more than a little touch of patriotism in Hongkong. Not so in Shanghai, a place of twenty different nationalities, already by 1870 one of the most cosmopolitan cities in the world, where no one was interested in who ran the telegraph so long as it was quick. There the Danes would have it — the shortness of the route.

Public response to China Submarine had been good in London; the shares had been taken up. There was no turning back. But unless Pender could make a deal with Great Northern in Copenhagen, the company risked being a financial failure. The slightest adverse turn of events, and it could be worse than a financial failure. China Submarine, by the pen of a clever newspaper editor, could all too easily be labelled a deception, after all.

The ensuing six months were among the most tremendous in John Pender's life, when he set in train, and began to see realized, an electrification of correspondence which changed the entire perception of distance between Europe, Asia, and Australia. His feat seemed almost too great for one man to accomplish. Yet he did accomplish it, and later in other oceans and continents as well, as the world of his day freely and with admiration acknowledged.

The first, and in some ways the most glorious moment in this sensational progress occurred on 23 June 1870, when the Penders gave a grand fête at their London home in Arlington Street to mark the inauguration of the telegraph to India. In splendour combined with good taste, which was Emma Pender's natural province, her husband received, on an instrument placed in the principal room of the house, the first direct telegraph message from Bombay, with the Prince of Wales standing beside him.

Yet underlying the months leading up to this splendid occasion, Pender had been engaged in desperately serious negotiations with the Danish company. At almost any cost he had to prevent a situation which in Britain could give rise to public concern. At some reasonable cost he had to ensure his own company's continuing stake in the telegraph traffic with China, which was quite clearly going to be opened by the Danes, although how they were going to do it remained a mystery which they were delightfully disinclined to reveal.

These were the inner circumstances in which the cable telegraph prepared to advance eastward from Galle — actually from Madras. The public in Britain knew nothing about it. The Danes, with their acute sense of situation, uttered not a word. As for Hongkong, it is doubtful whether people there had even heard of John Pender. Nor would they have been entertained to learn that his imperative aim was to reach Shanghai, not them.

By the latter part of May 1870 it was becoming evident — according to *The Times*, where someone had their eye on this peculiar China cable situation — that the Danish and British companies had come to some kind of agreement. A shared landing at Hongkong, it was suggested. Working 'in harmony' was mentioned. It was strangely vague.

Understandably so. The terms of the agreement concluded that month between John Pender and Great Northern could not be revealed to anybody, for one particular reason. The China Submarine company, despite what its prospectus had declared, was not going to reach Shanghai. Its cable would run from Singapore to Hongkong, and no further. In the negotiations, the Danes were unyielding on this. They themselves would lay their cable between Hongkong and Shanghai in six months' time; China Submarine would not reach the region for at least a year. For two companies to compete at the outset in an attempt to bring the cable telegraph to so problematic a country as China, the likelihood was that both would fail.

That fact, at that precise moment, before either of them had reached China, was beyond dispute. With the time factor added, it followed that there would be one cable from Hongkong to Shanghai, and it would be Danish.

Having conceded that point, which for John Pender was the harshest of all, he was left with the prospect that his China

Submarine company, though it might pay its way, was unlikely to yield any dividends. Thus, in the remainder of the negotiations, even the smallest points counted.

With the head of his Far Eastern cable at Hongkong, which could not be expected to be much of an earner, he had at least one consolation. The French Government, whose telegraph system connected with that of the British cable companies in the Mediterranean, and with whom there was a particularly cordial relationship, were anxious to see the telegraph extended to their possessions in Cochin-China. If John Pender ran his Singapore cable to Cap St Jacques, at the easternmost end of the Mekong delta, before taking it on to Hongkong, the French colonial authorities could run a land telegraph to Saigon, and this would give two sources of revenue on the Hongkong link. China, however, was where the large-scale traffic would be, through the Treaty Ports, as he saw it. Further than this, he saw in particular Japan.

Unimaginable in the East, with its enclosed civilizations locked resolutely in their incomprehensible past, into which even so modest and sensible an intrusion as the telegraph was viewed as an affront and a danger, Japan over the previous eighteen months had experienced internal change of an extraordinary nature.

This was the Meiji Restoration of 1868, when a powerful group of outward-looking Japanese, of a type which had been confined to their islands for two centuries by the Tokugawa shoguns, symbolically bore the Emperor from Kyoto to Yedo, releasing him from more than a thousand years of shadow rule, restoring him to his rightful position, in which he was to preside over Japan's assimilation of the science and learning of the West.

John Pender, with his Foreign Office connections, was as informed as anyone in the West could be concerning Japan. No one had more than a dim perception of what the changes there portended. Yet even if dimly, enough was known for one thing to be clear. It was more important in 1870 for the British cable system to reach Japan than it had been even in 1869.

Here the obstacle was presented that anything north of Shanghai — especially Japan — was part of the Danish grand design. If the British cable interest wished to extend northward from Shanghai, the Danes stated, it could be only along Danish cable lines. Pender made the counter-move that if the Danes wished to extend southward from Hongkong, this could be only along British cable lines. This was a misfire, in actual fact. Great Northern were not interested in extending south of Hongkong, and never had been. Nevertheless, from Pender's viewpoint, it would read well if China Submarine were ever called upon to justify its position before its shareholders.

This resolution to use each other's lines, to the north and to the south respectively, neither poaching on the other, was extended somewhat optimistically to include land as well as sea, and created two zones. In between these was an undefined region — the sector between Hongkong and Shanghai, to be connected by a Danish cable. This was declared a neutral area.

In regard to this area the Danes compromised, and John Pender was enabled potentially to strengthen China Submarine's uncertain financial outlook. The two companies were to share the profits of the Hongkong–Shanghai traffic equally, the Danish company to run the route at their sole expense. Both companies would jointly purchase or rent a suitable building in Hongkong for use as a joint station with separate receiving offices, enabling each to conduct their own business independently. In Shanghai and at places north of it, including Nagasaki, the Danish company would provide suitable accommodation and receiving offices for the British company. Before any division of the gross receipts on the Hongkong–Shanghai traffic was made, £15,000 was to be deducted for the northward extension from Shanghai to Nagasaki, the Danes agreeing to make this extension before the end of 1872, unless prevented by the Japanese Government or by *force majeure.*

The reason why the Danes made their profit-sharing compromise was that they realized that once their northward cables were laid, the trans-Russia traffic to Europe and America would far outweigh in value the traffic between Shanghai and Hongkong, and they could therefore afford the risk of compromising. Also, beyond the scope of immediate cable considerations, it was important that the British be reasonably satisfied. The British were the greatest power in the East; their Minister in Peking wielded more influence than any other diplomat there; their navy was the main deterrent to the endless piracy in the China Seas. All of this was important.

As with most good agreements — for this was what it was — neither party entirely liked it. Pender liked it a good deal less than the Danes did. When Britons came to operating it in China, it was found that they liked it even less than their chairman did, to the point of resentment in some instances.

This was the basis on which the submarine telegraph reached Hongkong.

The agreements made by both companies with the British Government were in essence identical. Each was granted the right to land a cable on Hongkong Island before a required date, priority to be given to all government messages subsequently transmitted. Her Majesty's Government had power to take possession of the cable in the event of an emergency, with reasonable compensation. The companies undertook not to dispose of the cable to any foreign government, or any other company, without British Government assent. Free enterprise not to be forgotten, the Colonial Office reserved 'full power to afford similar facilities to any other Company or Companies if at any time it may be expedient to do so'.

This side of the business had been formalized by exchange of letters with the Colonial Office, with Treasury and Foreign Office approval, when the Spanish Ambassador in London returned to the matter of the Philippine Islands cable concession for linking Manila with Singapore and Hongkong. The bureaucrats were right about the pleasant pre-Christmas delay. It had taken seven months to get an answer.

The Spanish Government's field of choice in this matter was peculiarly restricted. The Philippines was a country which in all but name was governed by priests and monks. Indeed, at times it was more explicit than this, as when between the departure of one Governor-General and the arrival of the next, the reins of government were held by the Archbishop.

Such overseas commerce as there was — it was mainly British, and under severe restrictions — was negligible. Cable traffic would be governmental and priestly (communication with the heads of the religious orders in Madrid), only a fraction of it commercial. Unless a forty-year monopoly concession was offered, it was unlikely that any company would be interested in making the cable link with the Islands, while the Government of the Philippines, which only just paid its way, could not possibly afford to lay a cable of its own, nor would even Spain entertain such a proposition.

The British Government, as one colonial power to another, professed to appreciate this predicament. They stressed that they were most anxious to see Spain's Philippine contact with Europe speeded up. If the monopoly clause could be removed, there would be no problem.

Spain as a result altered her arrangements with the two

Americans concerned, giving them sole rights to lay cables within two years, at the end of which, if the cables were not laid, the agreement would lapse.

The two Americans, as the Colonial Office had originally suspected, had been counting on their monopoly to secure the cable traffic of Singapore and Hongkong. They were clearly impecunious, and may even have been Britons posing as Americans in a Spanish colony, where no one would be any the wiser. With the now imminent advent of the British cable to these two places, they would have difficulty raising money for a cable system which, on its own, could scarcely be expected to run other than at a loss. In fact their scheme foundered on this, as the Foreign Office, behind a devious smile, quite expected it would.

The Philippines would thus remain dependent on the P. & O. weekly packet steamer between Manila and Hongkong. When the cable reached Hongkong, they would at least be able to send their telegraphic messages from there, which would be better than nothing.

Nor would it be long now.

Chapter 5

Danes at Deep Water Bay

HIS Danish Majesty's steam frigate *Tordenskjöld*, bearing the first length of the Shanghai cable, reached Hongkong on 16 September 1870.

Though the weather was still exceedingly hot and humid, in shadow there was a faint sensation of coolness, which a fortnight earlier there would not have been. It presaged the more equable seasons ahead: the serene and drier autumn when Europeans put on the kind of clothes they wore in Europe and looked more like their proper selves; the short sharp winter of January, with its starry nights and startlingly blue days; the grey moist spring.

Approached from the ocean on a fine day, Britain's most remote Asiatic possession was a towering grass-green mountain which was itself a complexity of peaks, bluffs, pinnacles, gaps, and staggeringly steep slopes, all of them green, with not a house to be seen anywhere. It was entering the great broad harbour on the other side of the mountain which took people's breath away, and swelled the patriotic heart of every Briton. For it was not simply impressive; it possessed beauty of that special kind which owes to an element of diversity: the beauty of many ships, the beauty of a dramatic mountain landscape, the beauty of fine buildings on a shore. Every person, no matter how different in mind, found something in it to admire.

On this side, facing the lesser hills of the Chinese mainland, the Hongkong mountain composed itself in a strangely satisfying majesty, a noble repose, of which the compelling feature was an upward linear sweep from shore to mountain-top, which artist after artist had been fascinated by, and had tried to convey correctly.

At the foot of this peak, which was over 1,800 feet high, was the city of Victoria, built along the shore, and rising up the lower mountainside slopes to as high as 300 feet in some places. Small by comparison with Calcutta or Madras, it was none the less imposing, and was rendered even more so because, after the first sight of a mountain with nothing but grass on it, the city came as such a complete surprise.

The colony's population numbered about 130,000, of whom 97 per cent were Chinese. Of the rest, about 1 per cent of the population were Macao Portuguese, a well-to-do professional and middle-class community, and 1 per cent were Indian. The remaining 1 per cent, which was where the non-Chinese wealth mainly lay, was composed of Britons, Continentals of all the countries from Scandinavia to Spain, Parsees, Jews, and Armenians, roughly in that order numerically, and, the Continental element excepted, nearly all of them British subjects.

Hongkong's first telegraph was a private one, set up in 1863 by Jardine, Matheson & Company, the greatest and most influential foreign firm in the Far East, and the commercial colossus of the China coast, founded by the opium-trade princes William Jardine and James Matheson long before Hongkong became British, or indeed had ever been heard of.

The firm had its headquarters at East Point, which jutted out into the harbour just to the west of Causeway Bay, where they had what amounted to a complete town, an irreproachably elegant one, its houses the final word in tropical architecture, sumptuous and comfortable, with lovely gardens. The town included a public market, known somewhat irreverently as Jardine's Bazaar, a slipway for careening and repairing vessels of the Jardine Matheson fleet, and of course stables and grazing for the Jardine Matheson racehorses.

In that they paid the peppercorn rent — one dollar a year — for the racecourse, which was in Happy Valley, almost next to them, they practically owned the racecourse as well.

The telegraph was introduced in order to attain quick connection between this formidable, albeit elegant, region of splendour, with the firm's town office in Victoria — a distance of about one-and-a-half miles.

Its poles and wires followed the shore line along Praya East, past suburban Wanchai and Spring Gardens. It then snaked between the naval and military establishments, and after further shoreline exposure reached the first building in Queen's Road Central. From there it was affixed from one building to the next, till it reached the office.

It is worth reflecting that when it was set up, there was no conceivable possibility of an international telegraph ever reaching Hongkong. No cable known to man could make the distance; the Danish activity in Asiatic Russia was entirely unknown, nor did it at that time include any idea of a juncture to China. Indeed, Jardine Matheson's having been known for more than a generation as the Princely Hong, their telegraph, while deemed intriguing, was seen as little more than another piece of princeliness. Office peons were quite good enough for most people when it came to sending messages.

Four years later it finally made an impression on the police chief, who persuaded the Government of the need for a police telegraph system. This would link the Central Police Station, next to Victoria Gaol, with the other town stations, and with the small police posts on the south side of the island. This was an area where miscreants took refuge, escaped, or came in, and was favoured by pirates of the more formidable sort — those with vessels carrying Western armament — for landing men ashore to go into town to pick up intelligence of shipping movements. Piracy was a well-organized business, and the scourge of the coast. A particular aim of the Captain Superintendent was to have telegraphic contact with these posts.

Because this was a government undertaking, local purchase of telegraphic equipment and stores (from Jardine Matheson's) was precluded; everything had to come from England, causing delay. Laying the lines was rendered difficult by the terrain, and by the sensible decision to go over the mountain rather than round it in one place, in order to give a shorter distance. Police clerks and some Indians in the force who could read and write — there were a few who could — were taught the Morse code and instructed in operating the telegraph. Early in 1869 the system was in operation.

When the *Tordenskjöld* arrived the following year, the existence of the police telegraph gave special pleasure in the Hongkong Secretariat. The coming of the new cable link with the world was one of the most important events in Hongkong's history. The opening of the Suez Canal had been another, but that was geographically far away. The coming of the cable was present and of immediate impact, comparable only with the moment when the P. & O. monthly mail steamer became fortnightly.

Apart from the fact that the cable would be going northward to Shanghai, when most Hongkong people would have preferred it to be going the other way, this was an auspicious occasion. Seen from the Hongkong Secretariat it had been auspicious from the first moment in London. The Secretary of

State for the Colonies was Lord Kimberley. He would not have written the dispatch regarding the reception which the *Tordenskjöld,* her officers, and all concerned with her undertaking, were to receive in Hongkong. He had certainly added to the draft, however, because by Colonial Office standards the dispatch was almost exuberant. Every facility the Hongkong Government had was to be made available, no impediment whatever to be placed in the way of the Danish operation. This was where the police telegraph came in.

The Danish cable was to be landed at Deep Water Bay. Thence, laid south and then north-east up the China coast, it would be immune to dragged anchors, whether off the recently opened Aberdeen Docks, or in Victoria Harbour. The Danish cable company naturally had to be in touch with Deep Water Bay, and their office was on the other side of the island, in Queen's Road Central. A road of sorts ran from there round the west side of the island, via Pokfulam to the Aberdeen Docks, but beyond that, to Deep Water Bay, there was no road. There was, however, a police post at Deep Water Bay, connected by telegraph with town. At a very modest charge, the Danish company would be able to use the government telegraph poles, and run a line of their own directly to their office. It gave a heart-warming sense of modernity to be able to make this offer, and to have it gratefully accepted.

A cable-house having been erected and fitted at Deep Water Bay, on 20 October the *Tordenskjöld* crew ran out cable to the shore, watched by the Colonial Secretary, Gardiner Austin — the Governor was away — and a gathering of officials and citizens. They had come by boat, the only practical way of getting there. In the shallower water the government steam launch, provided at no cost, played a useful part. Designed for anti-piracy purposes, she was an armed vessel of shallow draught, and very manœuvrable. For the last stretch the cable was manhandled just above water level, the head of the cable being borne up the beach to be deposited dry and safe in its home.

It had been an exhilarating and thought-provoking day. By the next mail steamer, Gardiner Austin wrote to Lord Kimberley to say that his Lordship's wishes had been carried out in full, drawing special attention to the handy use of the government telegraph poles. Two more Danish vessels had arrived by this time, and these 'will now be despatched with as little delay as possible to lay their respective portions of the line to Shanghai; and I shall at an early date be able to report to your Lordship the successful completion of the undertaking.'

The last of the three vessels departed on her strange and daunting mission — to bring the telegraph to a vast, enigmatic country which did not want it, and would probably not allow it in. Five months went by without any certain news, and certainly no telegraph service.

During this time — actually in December — John Pender's British-Indian Extension company, acting with that speed and sureness which themselves provoke admiration, brought the cable eastward to Penang and Singapore. One is left astonished by how much of the world was brought within the range of the telegraph in the year 1870, really thanks to one man.

His ideas about making the link with Australia had matured to the extent of the formation of yet another company. Already a cable had been laid from Singapore to Java, where the Dutch overland system bore messages to the eastern end of that island. Another cable from there to Timor; then Australia would be next. How the cable would be laid depended on the Australian colonists bringing their overland telegraph to the north of their country, and where they would bring it to. It would take a year or so.

As things stood in Hongkong, a telegram could reach London in five days, if sent by a ship with direct sailing to Singapore.

Then, in the Shanghai newspapers of 27 April 1871, copied ten days later in the Hongkong newspapers and in the *Japan Times* of Yokohama, the Great Northern Telegraph China and Japan Extension Company announced that with immediate effect it would forward telegrams from Shanghai, via Singapore, to all parts of the world where there was telegraphic communication by established lines. Between Hongkong and Singapore, telegrams would be sent by steamer, free of charge. Danish Consulates in Japan — there were three — would forward telegrams to the company in Shanghai, for transmission to all parts of the world similarly.

How had they done it?

Chapter 6

The Shanghai Taotai's Silence

ONE of the extraordinary things about the cable landing at Deep Water Bay in 1870 was that people in Hongkong were unaware of there being any political difficulty about it. Here the fact was that the Hongkong public knew nothing about China's refusal to allow the overland telegraph into their country — from Kyakhta, which no one had ever heard of anyway.

In the great extension of the telegraph with which we are dealing, the public knew when something actually happened. Of how anything had come to happen, the public, in whatever country, knew almost nothing.

Before there was any talk about a telegraph to Shanghai, it was felt in Hongkong that there might be trouble with China on such a matter, the telegraph being new, foreign, unknown, and — worst of all — electric. This meant that it was the purveyor of an uncontrollable spirit, most vividly demonstrated in lightning, which was manifestly malevolent in that it caused instant death. It was because the wires conveyed messages by harnessed lightning that the telegraph was considered contrary to nature.

But when the *Tordenskjöld* arrived, and when her officers and the civilian personnel of the Great Northern Telegraph went confidently and serenely about their business, and particularly when local guests were invited aboard to noonday parties, at which His Danish Majesty's naval officers gave evidence of their internationally unrivalled imperviousness to the effects of schnapps — a distinction perceived by guests only when they themselves came to, which was often the next day — it was not unsurprisingly assumed that China must have given them permission to bring the cable to Shanghai. As when anything telegraphic actually happened, nobody bothered to enquire.

In reality, no permission had been given, nor had either the Danish Government or Great Northern asked for any. A request, being almost certain to be met by a refusal, was to be avoided.

23 June 1870. The first telegram from Bombay is read out to the Prince of Wales at John and Emma Pender's London home; extreme right is Ferdinand de Lesseps. *Below,* the modest Bombay office from which the telegram was sent.

Above, Danish routine inspection of the overland telegraph in Eastern Siberia, about 1890.

Opposite, at Hongkong Peak signal station in the 1890s flags announced the approach of mail and cargo ships, and their nationality. The Jardine Matheson stronghold at East Point juts out into the harbour, and the causeway has been built across Causeway Bay.

Below, a camel caravan approaching Peking. The caravan route through Mongolia was the best route for the telegraph, but the Chinese Imperial Government would not allow it.

1870, the year the telegraph arrived. *Above*, the Hongkong waterfront, not meeting expectations; *below*, the Bund at Shanghai, where the real business was.

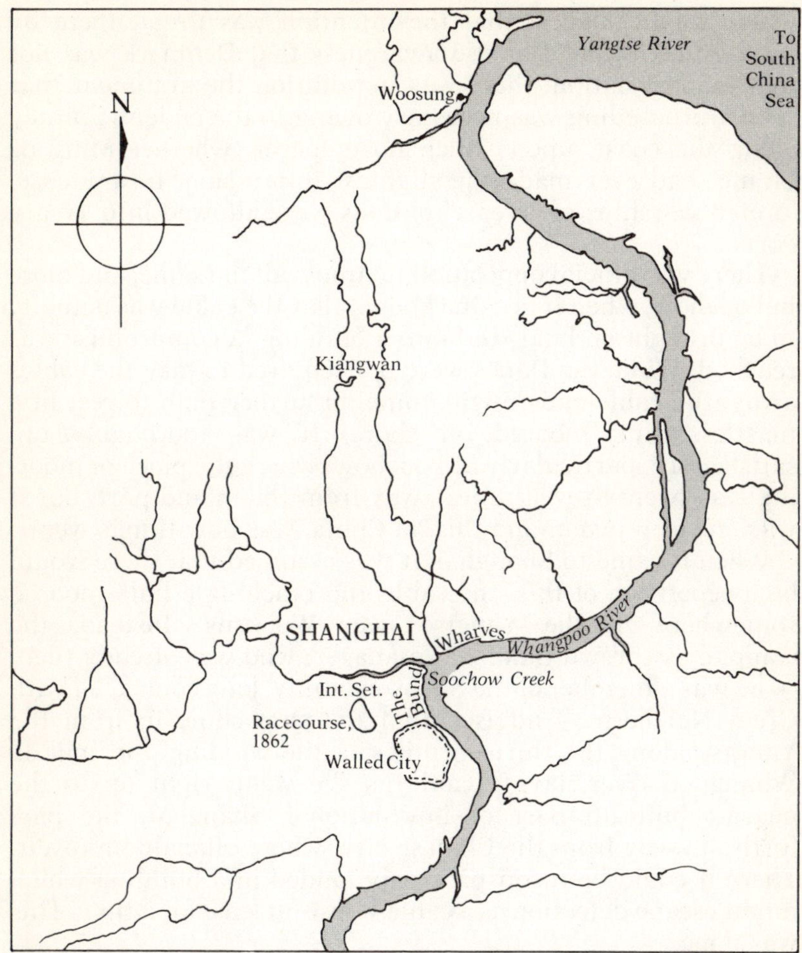

SHANGHAI AT THE COMING OF THE CABLE TELEGRAPH, 1871

They had already been refused once, in the matter of the overland telegraph, and had in Peking's view shown commendable obedience in not pursuing the course further. They were in comparatively good odour with Peking, in other words. This time the idea was to proceed by action first, in suchwise as to place no Chinese authority in the embarrassing position of having to make a choice. The Danes, with folk memories of the old Canton days, long before there were any Treaty Ports, were old hands at this kind of thing.

The cable, it will have been noted, had come in a warship,

which might suggest that the intention was to get there by guns. Allowing for Chinese awareness that Denmark was not that kind of nation, there was in addition the argument that such a proceeding was necessary owing to the endemic piracy along the coast, upon which no emperor, whether Ming or Ch'ing, had ever made the slightest impression. In any case, foreign warships of 'Treaty' nations were allowed into Treaty Ports.

There was official opposition in Amoy, all the same, and more in Foochow, when it became known that the cable was actually to be brought on land, and into a building. A compromise was reached. While the Danes were not required to take the cables away, the cable ends might come no further than to rest in a mastless hulk moored off shore. It was thoroughly unsatisfactory, particularly at Foochow, where the place of mooring was twenty-seven miles away from the inland port, but it was one step taken; and this, in China, was how things went.

When it came to Shanghai, it was assumed that there would be a repetition of this, the cable end reaching a hulk moored somewhere in the Yangtse near Woosung. Instead, the company's shrewd Shanghai manager, who was already there — he was a member of the Suenson family, long connected with Great Northern — advised that the ship come in from the Yangtse along the thirteen miles of the winding and muddy Whangpoo river, laying cable as she went, right up to the nearest built-up part of international Shanghai, the part furthest away from the Chinese city, where officialdom dwelt. There it could be inconspicuously landed in a building which might escape detection as a cable-house, at least for a time. This was done.

For a man new to China, Suenson's advice would seem bold to the point of the foolhardy. His confidence reposed in having tested his views against a foreign source of experience greater than any in China — that of Jardine, Matheson & Company.

Most people had forgotten about it, and many had never known, but for years Jardine Matheson's had been the Danish Consulate in China. It dated from around 1820, in the days when the East India Company obliged all Britons, in accordance with Chinese rules, to leave Canton at the end of each year's trading season and go down to Macao, an annual upheaval which private British traders complained about greatly. To circumvent these orders James Matheson, then in his early twenties, the most dashing and successful trader on the coast, sought and was awarded the post of Danish Consul, a position which conferred upon him *de facto* Danish nationality, rendering it impossible for the East India Company to interfere with his movements. After a few years, he and William Jardine joined forces under the flag of Denmark, and whoever was head of the firm and resident in China was the Consul. For years, wherever the flag of Denmark was to be seen in China, it was a sign that the building it flew from was Jardine Matheson's. It no longer flew, but it was there that Suenson knew there was a welcome.

The Great Northern Telegraph, it will be remembered, had been denied the assistance of British Consuls in China. The Princely Hong's influence was such that its assistance was nearly as good, and its advice probably better. One cannot help wondering what John Pender would have made of it, if he had known, after his difficult Danish negotiations. China was a very strange place.

None were keener than Jardine Matheson's to have telegraphic contact with Hongkong. The volume of their Shanghai business was many times larger than that of their Hongkong headquarters; quick intimation and response were vital to them. The presence of the cable-head in that inconspicuous building on the wrong side of Soochow Creek would inevitably become known to His Excellency the Taotai of Shanghai. It may have been known within hours of its being landed. Shanghai was a busy place, though, and the building was just far enough away for the Taotai not to need to know about what was inside it, unless someone did something stupid, when it might become necessary for him to know — which would happen very fast. Shanghai was another very strange place, in some ways even stranger than the country of which it was part.

The Taotai of Amoy and the Viceroy at Foochow had struck an attitude when the cable came, and imposed limits, as was their minimal bounden duty. The Taotai of Shanghai struck no attitude, uttered not a sound, moved not a soldier. Provided absolute discretion was observed by all concerned with it, the telegraph was there. It had reached China.

Foreigners would use it first, and a notice to this effect in foreign newspapers would not be observed. Being in foreign language, such a notice was one which the Taotai was not required to observe, it being beneath the dignity of a high official to give attention to anything in a foreign written language. It was a situation which led onwards. Once a system of transmitting Chinese words by telegraph was worked out, which would not take long, Shanghai Chinese might start using it. With no permission given or asked for, it would become part of the scenery.

Nobody did anything stupid.

Chapter 7

A Cable to Queen Victoria

THE China Submarine Telegraph Company's cable reached Hongkong nearly two months earlier than expected. John Pender had undertaken to the British Government to have the cable laid, and the service in operation, by the end of July 1871. In fact, the British cable ship, having started at Singapore, and having already landed a line at Cap St Jacques in Cochin-China, reached Hongkong in the last days of May.

The cable was landed in the Pokfulam area on the west side of the island, directionally the most convenient side, once again out of range of dragged anchors, at a small inlet which quickly became known as Telegraph Bay. As with the Danish company before, there was the special convenience that, at a very modest charge, the British company could use the government telegraph poles to run a line to the Queen's Road office which they shared with Great Northern. The road between Pokfulam and Victoria was not bad, but it was not a carriage road.

Despite the comparative nearness to town, little ceremony attended the landing. It was very hot and pouring with rain, day in, day out. Ceremony was confined to the opening exchanges of telegrams with London.

The Governor was away — we seem to have heard this before. By some perversity of chance, the Governor, whichever one it was, always seemed to be away when anything telegraphically important took place. On this occasion General Whitfield, in charge of the forces, was standing in as Lieutenant-Governor. To him, on 4 June 1871, John Pender, from his office in the City of London, addressed a telegram:

> Completion of Submarine Telegraph between Singapore, Straits, and Hongkong, brings the Chinese Empire into immediate connection with India and Europe.
>
> I congratulate you on the important result, and hope it may prove equally advantageous to all the countries so connected, in promoting friendly relations and commercial intercourse.

It reflected how John Pender viewed the significance of his enterprise. It reflected the truth. Whitfield replied:

> I thank you for your courteous message, and cordially congratulate you and your able staff on the completion of your great telegraphic enterprise, uniting with you in the hope that ...

The following day, 5 June, Pender sent a similar message to the chairman of the Hongkong Chamber of Commerce. Before sending this, he worked out what the time would be in Hongkong, directing that the operator give the hour and minute of dispatch in Hongkong time. The telegram was sent at 1.07 p.m.; it was received in Hongkong at 2 p.m. precisely. Fifty-three minutes. Only seven years before, the fastest message between London and Hongkong took forty days.

The message had come on a cleared line, of course. Ordinary messages would take several hours. Nevertheless, it was an utter marvel, diminished only — as the *Hongkong Daily Press* pointed out — by the 'astonishing complacency' with which the public regarded it. The editor found himself positively having to explain to them what a marvel it was.

The editor, of course, was thinking in terms of the 1 per cent of the population who might be expected to read his newspaper. The Chinese community needed no such explanations. By Chinese in international commerce, the telegraph was viewed as a miraculous and wonderful improvement.

The contrast between this reaction and Chinese official opposition to the telegraph is striking. The attitude of the Ch'ing court at Peking was clear; there lay the core of opposition. The attitude of the gentry was a surmise — officialdom's surmise; it had not been put to the test. As to the populace at large, a great many would take their line from the gentry. Beyond this all that could be said was that if wires were laid in China they would probably be stolen.

Hongkong Chinese business men were not of the gentry. No business man was of the gentry, commerce ranking in China among the lowest of all human occupations. Added to this, although pictures of them with their pigtails and little caps make them seem picturesquely antique, they were in fact very modern. For instance, while convinced that anything electric contained a 'spirit', they did not think that a telegraph wire transmitting messages by harnessed lightning was dangerous. So modern were they that they often found themselves talking down to their relatives in Canton, and waking them up in the process. Hongkong was the seed-bed of modern China, and Hongkong Chinese reaction to the telegraph is a small example of this.

Europeans, unless they were in shipping, knew very little about Chinese international commerce, because it was commerce with Chinese communities overseas. Cantonese conducted a lively export trade to California, providing Chinese there with the wherewithal of Chinese living — everything one can think of: dried and preserved foodstuffs, clothing, furniture, roofing tiles, slippers, spittoons, name it and it came from Hongkong. Merchants in this diverse business could now send a message to San Francisco and get an answer in three days, which verged on the unbelievable.

Another and even larger line of business was with the Nanyang, the South Seas, in particular with Singapore, Penang, and Surabaya. This was trade in rice, medicines, and tropical products, and was conducted by the Teochew-speaking community. There had for long been Teochew communities in the Nanyang. When their home town, Swatow in eastern Kwangtung, became a Treaty Port (1860), these communities enlarged, while in Hongkong there was a Swatow civil invasion which led to the conquest of Sheungwan, just west-of-centre in the city of Victoria, a zone which became and remained a Teochew-speaking stronghold. For merchants there, to be able to send a message to Singapore in the morning, and have an answer perhaps in the afternoon, was like something out of a dream.

Cantonese did not like Teochew people, nor could they understand a word of their language. Teochew is more than a dialect, being as different from Cantonese as English is from Romanian. An astonishing feature of the coming of the telegraph was that the leading merchants of both communities overlooked their differences, being united in their desire to send a message of gratitude for this new and wonderful manifestation of progress.

General Whitfield had already exchanged congratulatory telegrams with Lord Kimberley. The leading Chinese merchants, an *ad hoc* committee of them, did not know who Lord Kimberley was. In their way of thinking, there was only one person to whom gratitute could properly be expressed. That person was Queen Victoria.

Subsequent generations, poring over Hongkong's past, were to find this laughable. In reality, prior to the Chinese Revolution of 1911, it was entirely in character, the world of those times being an imperial world; and there was a special and intriguing quality in this particular instance. Had this event been taking place in Canton, the merchants would have addressed the Viceroy, and not the Emperor, who was impersonally distant and could not be addressed. In Hongkong it was different. Something in the nature of British colonial government made Queen Victoria seem nearer than the Emperor of China, despite the opposite being the case. Something in the Queen's personality too, even in her long years of widowhood and deep mourning, conveyed itself mysteriously to all lands and races, making her seem nearer than she was.

When the deputation of merchants, having been rebuffed at the Secretariat and told to send their telegram to the Colonial Office, called at Government House to make it clear that they wished to address the Queen in person, the situation, far from being laughable, was peculiarly disconcerting, the General having somehow to avoid telling them that not even he could send a message directly to the Queen. He insisted that the telegram be sent to the Colonial Office, though in the end conceding that a request for transmission to the Queen might be included.

Upon this the deputation withdrew, went down the hill, and had the telegram rewritten by someone with a very good knowledge of English. Even if obediently dispatched to the Colonial Office, it was so worded as to be addressed uniquely to a person who was not at the Colonial Office, but at Windsor Castle. When they returned with this to Government House the following day — three days were consumed in this palaver — their cleverness was noted, and it was so clever that nothing more could be said. The telegram, sent that day, 12 June 1871, read:

> Respectfully offer congratulations to Her Majesty on completion of telegraph between London and China. May civilization advance. Long live the Queen. Peace, prosperity. Chinese Mercantile Community Hongkong.

At the Colonial Office the received message started in the morning at ground level, abode of short tempers. 'We have congratulated the Colony by telegram to the Governor', that level minuted irritably. It rose to higher, purer air. 'But I think it should be acknowledged.' This was to Lord Kimberley, higher still, and it reached him before he left for luncheon.

Nothing more happened till late in the afternoon of the next day. Then Kimberley minuted down:

Answer. Have laid before the Queen and am commanded to convey satisfaction with which Her Majesty has received this intelligence and the loyal message of the Chinese merchants.

As those few who had access to Queen Victoria during those years knew, messages of this kind from her subject peoples of other races gave her a particular happiness.

Over the following months, with the Danish cable securely ensconced in Shanghai — the Taotai's marvellous silence — things moved with marvellous speed. Great Northern had undertaken to John Pender to have their cable through to Nagasaki by the end of 1872, unless obstacles arose from the Japanese Government.

No obstacle arose. The Meiji Restoration had altered many things in that country. On 14 August 1871, more than sixteen months in advance of the deadline date, and only two months after the British cable service opened in Hongkong, it was announced in London that telegraphic messages from London to Nagasaki could be sent at the rate of 20 words for £9.5s.

The announcement was made by China Submarine, tactfully giving the British people the impression that it was a British line, whereas from Hongkong onwards the cable was of course Danish. Messages were to be marked 'via Falmouth', the starting point of Pender's truly immense imperial line, which even a century later inspires awe of the man's achievement, as also of Tietgen, whose visionary intention was about to be fulfilled.

The Danish cable was laid from Nagasaki to Vladivostok. On 11 January 1872, in London, the Great Northern Telegraph Company announced that it was forwarding messages via Russia for Hongkong, Shanghai, and Nagasaki, at a uniform rate of 20 words for £4.6s. Messages were to be marked 'via Russia'.

The difference in price — less than half the price of a Pender telegram — was immediately spotted at the Colonial Office.

'I think we should avail ourselves of this line whenever opportunity occurs', the ground level minuted.

'Certainly', came an instant reply from higher level.

There was no special support for British undertakings in that part of Whitehall, apparently. As a cynic in Hongkong observed, however, the real question was whether the Danish rate was so low it would force the British rate down, or whether the latter was so high it would allow the former to go up.

Soon after this, John Pender merged his three companies operating east of Madras — the Indian extension, the Australian, and China Submarine — and in May 1873 formed the Eastern Extension Australasia and China Telegraph Company, usually known simply as Eastern Extension.

In the evenings it advertised its presence in Hongkong in a manner which was long to be remembered. Having some extra current to play with, the engineers fixed up an arc lamp over the front entrance. In the attractively gaslit street, it beamed like a lighthouse.

PART 2

A Study in Slow Motion

In his after years Sir John Pope Hennessy was fond of talking about the 'cabals' formed against him in Hongkong, and it is true that he had made many enemies who longed to see him leave the Colony 'under a cloud'.

James Pope-Hennessy,
of his grandfather,
Verandah, 1964

Previous page: A demonstration of Graham Bell's 'butterstamp' telephone was made in Hongkong in 1877, a year after Bell had first succeeded in transmitting speech from one room to another.

Chapter 8

Court Circles

THE inventor of the telephone was an Edinburgh-born Scottish doctor, Alexander Graham Bell, educated and trained in the Universities of Edinburgh and London. He was a specialist in teaching deaf-mutes to speak. At the time of making his invention he was Professor of Vocal Physiology in the University of Boston, Massachusetts. He patented the invention in the United States in 1876, when he was twenty-nine years old.

When he demonstrated it, at a scientific and industrial exhibition held in Washington that year to mark the centenary of the Declaration of American Independence, public reaction to it disconcerted him. Americans, a literal and down-to-earth people, envisualized the telephone exactly as it had been demonstrated to them: as a means of conversing across a short distance with someone who cannot be seen and who is otherwise inaudible. It was a short-distance instrument.

In vain Graham Bell lectured to them. It was a long-distance instrument. 'No limit has been discovered to the distance over which sound can be transmitted', he told the New York Society of Arts in 1877. Nor was the telephone a matter of speaking solely between two given places. It might seem fantastic, but a time would come when there would be wire-joining exchanges by means of which any person with a telephone could speak to any other such person, in any home or office. There could one day be such an exchange for the entire United States.

His words fell on deaf ears. Americans simply could not believe it.

Forty years before, when the telegraph was invented, it had taken seven years for its various inventors — it was invented in several places at almost the same time — to impress upon responsible opinion an appreciation of its value and usefulness.

The telephone, despite being mistakenly seen as a short-distance instrument, had a reception markedly different — an immediate reception. Telephones were in use in the Far East sixteen months after Bell's demonstration.

Just before the first of them were installed, a telephone made in Hongkong was demonstrated. Its maker, João Maria da Silva — he was thirty-nine at this time — was the Audit and Reference Clerk in the Colonial Treasury. Of well-to-do Macao parentage, he was a child prodigy in scientific knowledge. Educated in Calcutta, where he excelled in chemistry, he joined the Hongkong Government at the age of twenty-one, and spent his entire working life as a civil servant. Assured thus of a reasonable income, he was free to devote himself to his extraordinary range of scientific interests: chemistry, electricity, botany, animal magnetism, and photography, of which he was an advanced exponent. In 1873, in addition to his work at the Treasury, he became the official Electrician and Inspector of Telegraphs. His merits were recognized at a distinguished level. He was a member of the Institute of Electrical Engineers in London, and of the Geographic Society of Lisbon, and in his fiftieth year was honoured by the Portuguese Crown, becoming a knight of the Order of Christ.

Learning of Bell's invention from scientific journals to which he subscribed, he obtained the required materials, some of which had to be sent for from Europe. He then made two telephones, and in November 1877 demonstrated them at Government House on a wire between two fairly distant rooms, with the Governor, Sir John Pope Hennessy, the Admiral, and numerous officials present.

The Governor had a lot of time for Macao Portuguese. Most Britons looked down on them. It is to be suspected that he was more diverted by the social discomfiture the occasion caused the Admiral and the others than by the telephone itself, in which he showed no further interest.

In Shanghai a week or two later — in December 1877 — Werner Siemens' famous electrical firm, which had pioneered the telegraph in Europe and West Asia, installed telephones for the China Merchants Steam Navigation Company, connecting their town office with their wharves, which were three miles down-river. Tokyo had its first telephones at about this time, so this may not have been the first working telephone in Asia, though it was certainly the first in China. Even more significant, the company which owned it was Chinese.

They were only just first. Jardine Matheson's had telephones in Shanghai a few days later, in January 1878.

Many people, however, were still not quite sure what a telephone was or did. 'Professor Bell asserts that the *quality* of the voice is conveyed by the Telephone, so that the person speaking can be recognized at the other end of the line', a local newspaper explained. Two Britons organized a demonstration at the Royal Asiatic Society's rooms in Shanghai. The aim was to talk to a barrister whose chambers were across the road. Unfortunately, quick as lightning, a Chinese cut and stole the wire, so the demonstration had to be inside the building. It was a slight disappointment to some who had not realized that only the person holding the ear-piece could hear the voice at the other end.

A more informative demonstration was given in Hongkong on 15 February 1878, when by popular request J. M. da Silva invited a group of men, including gentlemen of the press, to his large and comfortable home in Old Bailey Street, where tests were carried out using his own telephones. In the drawing room 'wires were laid to a magnetic battery, one communicating with an upper room, the other, or earth wire, with the gas pipe in the same room'. With the drawing room doors closed, conversation was held with people upstairs, various questions and answers being distinctly heard. When someone sang, however, though the tune — it was *Home, Sweet Home* — was clear, the words were not. (One hesitates to say so, but this might have been due to the singer.) A distance test was then done, along a wire laid over the rooftops to avoid theft, to an adjacent house facing the next street, a distance of fifty yards. 'In that test, the words of a conversation were plainly heard, but a piano, though distinct, appeared to come from a distance.'

Charles Vandeleur Creagh, the acting police chief, had been impressed by J. M. da Silva's demonstration at Government House, which he had attended in his capacity as temporary aide-de-camp to the Governor. Creagh saw the special advantage the telephone would have for a police force which was largely illiterate. Only a very few, those who could read and write, could communicate by telegraph from one police station to another. Messages had to be written before being sent. Since the telephone conveyed actual speech, any constable would be able to use it.

He mentioned the subject in his annual report for 1877. The moment this was published (in the spring of the following year), the Great Northern manager in China, another member of the Suenson family, was on to him. There, however, things broke down. It was the old problem of local purchase; a police telephone system being a government concern, all the equipment would have to come from England, where, in contrast to other European countries, there was only a tepid interest in the telephone. All the Shanghai telephones in operation were from Germany.

It should be possible, Creagh considered, to ask the Colonial Office to give unqualified permission for a police telephone system — some years later they actually did this — if the Governor would agree to put it to them. At this point however, Deane, Creagh's superior, returned from leave, and did not take kindly to his deputy having gone into print on so unusual an innovation as the telephone without consulting him.

There was another angle to this. Creagh was on very friendly terms with the Pope Hennessys, Sir John and his beautiful Eurasian wife Kitty. Police chiefs would be less than human if they did not regard with an uncongenial eye deputies who move in Court circles. There was the added difficulty that Vandeleur Creagh was a cut above Deane socially. He was also a first-class officer, held in high esteem by the public.

Creagh could easily have prompted the Governor regarding a police telephone, except that Sir John was essentially political in character. As his lack of response to Silva's demonstration showed, he was simply not interested in mechanical ingenuities — unless there was a political slant to them. In this instance there was none. Were he to raise the matter with Deane, as Creagh knew, it would be so out of character that Deane would immediately detect the identity of the prompter, rendering that eye on Vandeleur Creagh even more uncongenial than it already was.

This was the reason why Hongkong did not have a police telephone — or indeed, in view of the Governor's and everyone's indifference, any telephones at all — till years later than most of the modern cities of the Orient. In a surprisingly short time Tokyo, Shanghai, Singapore, Madras, Calcutta, Batavia (Jakarta), and Saigon, all had telephones, private or governmental. There were no central exchanges, of course, but some of the systems were quite elaborate. In Singapore there were telephones linking Government House, the Chief Secretary's office, several government departments, police stations, and the docks. Hongkong had none.

To ask how so trivial a situation could produce an impasse lasting several years, the short answer is that Hongkong was

The *Tordenskjöld*, bearing the first cable to a China unwilling to let it in. *Below left*, T. F. Tietgen, the Copenhagen banker who masterminded the venture; *below right*, Edouard Suenson, who landed the cable at Shanghai.

Hongkong, 1870: the first cable house at Deep Water Bay, *above*, with the Danish vessels *Tordenskjöld* and *Cella* at anchor. The overseas telegraph passed over Wong Nei Chong Gap, *below*, using existing police telegraph poles, seen on the right; in the distance Kowloon and, to the left of it, Stonecutters Island.

Hongkong viewed from East Point in 1870. *Below*, the British cable vessel *Kangaroo*, owned by China Submarine Telegraph Co., which made the Singapore–Saigon–Hongkong cable link in 1871.

On Queen's Road Central, *above*, the telegraph office's lamp, seen on the right, was until 1890 the city's one and only electric light. *Below*, inside the office, about 1890.

that kind of place. It also needs to be borne in mind that the British civilians, including women and children, numbered little more than 800 souls, and this small community was divided into classes, categories, and cliques which did not mix with each other socially. If anything went awry in any segment of such a society, it could usually be rectified only by death or departure.

Creagh eventually settled for departure. He had, before his transference to the police, already distinguished himself in the civil administration. Finding life insupportable under Deane, and Sir John and Kitty knowing this, silent arrangements were made from Government House for his transference to Perak, in the Federated Malay States, in the capacity of Assistant Resident, an administrative post of great responsibility. There his superior would be one of the most eminent Englishmen in the East, a name to be famous in Malayan history, Sir Hugh Low.

He was Kitty's father.

The moment Deane knew that this move was fairly certain to take place, the introduction of the telephone into Hongkong showed signs of becoming peculiarly easier.

Chapter 9

The Great Viceroy

DURING these years of telephonic impasse in Hongkong, a momentous development took place in North China.

Great Northern had not succeeded in bringing the telegraph to Tientsin, a place of importance in respect of the diplomatic telegraph traffic of Peking, only eighty miles away. Tientsin was doubly important by this time, in that since becoming a Treaty Port in 1861 it had soared into commercial prominence.

The Great Northern cable actually rested in a hulk moored off Taku Bar at the mouth of the Peiho. The distance from there to Tientsin, as the crow flies, was about twenty-five miles; but the Peiho is an impossibly winding river so the cable distance would be more like sixty miles. A river, as the company knew from its experience of the Whangpoo at Shanghai, was extremely vulnerable to cable theft. Time and again, in the first years of the telegraph, the service had been interrupted by thefts of cable in the Whangpoo, and the company were not prepared to take the greater risk of a cable up the Peiho. As was widely said at the time, in no part of China were poles and wires more needed than over those twenty-five miles between Taku and Tientsin.

It happened that there was a Chinese who agreed with this. Li Hung-chang, by illustrious service to the State, had risen to the highest of ranks, being at this time Viceroy of the metropolitan province of Chihli, and Grand Secretary of the Empire. In the course of his career he had seen more of foreigners than most Chinese high officers usually did. He had known General Charles Gordon — 'Gordon of Khartoum' or 'Chinese Gordon' — when, Li himself being of the rank of Futai in Kiangsu province, Gordon was given command of the Ever-Victorious Army, and had finally stamped out the Taiping Rebellion.

This, in fact, was the event which brought Li Hung-chang into commanding prominence. Demanding the submission of the Taiping 'princes', who between them had brought death by massacre to more than twenty million human beings, he promised them their lives, received their peaceful submission, judiciously overlooked his promise, and had every one of them executed. The Emperor confirmed him in his rank, bestowed on him the Yellow Jacket, and designated him Junior Guardian to the Heir Apparent, an honorary post with no duties, but which indicated in Court circles that he was selected for the highest preferment.

Li Hung-chang desired to see China modernized in certain ways. He wished it to have a railway and telegraph system, a modern mercantile marine, and financial unification. But he had to proceed carefully. In the highest ranks of government, which he shortly entered, there was total opposition to such ideas, and no matter how powerful a man was, Peking was a place of secret, silent, interminable feud.

He resolved to start on ships, a matter which, being dealt with in Shanghai, would be furthest away from the Emperor, arousing minimal attention and, it was to be hoped, no alarm. Getting various Chinese business men together — totally beneath the dignity of an official of his rank, and utterly astonishing — he in 1874 brought about the founding of the China Merchants Steam Navigation Company.

China Merchants, as they were usually called, did not build ships. They bought Western-style ships which had seen better days, repaired, refitted, and ran them. This was China's first modern company, a portent indeed. Three years later, as already observed, they owned the first telephones.

A year before this, Li Hung-chang had obtained the Emperor's consent in principle to introducing the telegraph into China. Foreigners heard of this by rumour, and when nothing came of it for several years, it was assumed that the rumour was false. In fact, Li Hung-chang was biding his time. Consent in principle could be withdrawn at the flick of anyone's fan. The Viceroy of Chihli needed to exercise patience, allowing consent in principle to settle into a mould.

In 1879 he deemed the time ripe, and picked upon the exact route on which to start — Taku to Tientsin. Once again he worked indirectly through merchants. These made contact with Great Northern. With Danish technical advice and under Danish supervision, China's first land telegraph was installed, Great Northern bringing their cable ashore at Taku into a Chinese cable-house. Although commercially financed and run, the land telegraph was under government control. The Viceroy was astute.

Under his orders, be it noted, the line ran solely to Tientsin, no further. If it were extended to Peking, he calculated, opposition would be overwhelming.

A Chinese land telegraph from Tientsin to Shanghai followed, opened in 1881. A telegraph line was at the same time run beside a road from Shanghai to Woosung on the Yangtse, and Great Northern landed their cable there, terminating the saga of cable theft on the Whangpoo. A line was opened between Shanghai and Nanking in 1882.

In that same year Li Hung-chang established the Chinese Telegraph Administration, a commercial company under government control, all existing commercial lines being transferred to it. As Director-General he appointed a much younger man whom he knew, Sheng Huang-hsün, a sound and capable administrator, who was much of one mind with him.

Expressed thus, it sounds as if things went smoothly. In fact, behind the scenes, Li Hung-chang was engaged in a major power struggle. The telegraph — as with railways and financial reform — was resisted because of its unifying effect. Officials high and low were opposed to it, seeing it as a threat to provincial autonomy and their virtual independence. Li mastered it — subtly, by sure steps. Despite Court misgivings, Peking was at last connected in 1884, with sighs of relief from the Diplomatic Corps, who had had to wait for it for twenty years.

Down in Hongkong meanwhile, Spain had been granted cable landing rights, in 1879. This was the outcome of a formula agreed between Madrid and London. The Philippine Islands had simply no alternative to monopoly operation if they were ever to be connected with the rest of the world. The formula was expressed in a letter to the Foreign Office from the Spanish Ambassador in London: 'The Spanish Government has decided to link the Philippine Islands by cable with the coast of China.' On this basis the British governmental ruling against monopolies was waived. Siemens, through their Indo-European Telegraph Company, secured the contract, and Eastern Extension laid the cable, which opened for service in 1880.

This led to an improvement in weather reporting, which by comparison with Europe was almost non-existent in the Far East, bringing the Manila Observatory, run by Jesuits, in touch with the Observatory in Hongkong, a factor of special value during the typhoon season. If a typhoon is going to affect Hongkong, it has nearly always affected the Philippines first;

typhoons, arising in the Pacific, move west. A telegram from the Manila Observatory concerning a typhoon gave Hongkong three or four days' advance warning of it.

Enston Squier, the Eastern Extension manager in Hongkong, entirely on his own initiative, obtained the approval of his (Pender) directors in London, contacted both Observatories, and in March 1882 arranged for a daily exchange of weather reports between Hongkong and Manila, transmitted by Eastern Extension free of charge.

A regional touch had come into Hongkong's electric contacts. Till then, in the electrical sense, it had been a dot on the map, connected with the great world yet as lonely as the arc lamp over the office door, the only electric light in town.

A current of regional warmth now reached Hongkong from the Philippines. Next would be the surely inevitable southward extension of the Chinese inland telegraph to Canton, regionally warmer still, the city which nearly every Hongkong Chinese business man wished to be in touch with.

Chapter 10

A Foreign Company's Fate

GERMANY, then the Scandinavian countries, and Switzerland were the earliest places where the long-distance possibilities of the telephone were perceived, leading to long-distance technical development. In Britain, public interest in the telephone was tepid, and for the first few years there was little telephonic activity. If any view of it was held at all, it was the short-distance view.

In the midst of this was John Pender, the most consequential man in the communications business. His view of the telephone was a special one. While entirely taking the short-distance view, he needed to decide what he could do with it. As a first measure, he acquired rights in Bell telephones.

The High Court had ruled that the telephone was a telegraph in the meaning of the Telegraph Acts. This decision meant that any telephone service within the British Isles would be within the monopoly of the Post Office.

Pender then made a bold and commercially brilliant move. The Government of India had ruled from the start that the telegraph was a government concern. Before they woke up to the British High Court's ruling on the telephone, and followed suit, Pender decided to introduce that instrument in India on a commercial basis, his own.

Under the aegis of the Oriental Telephone Company, which he set up in London in 1880, telephone companies were to be founded and systems installed from central exchanges in India and at points East. This was the short-distance telephone. Each major city in India would have its own Pender subsidiary company and system, giving service to the city and its environs. Between one city and another, one would use the government telegraph. The entire telephone operation was seen by Pender as an adjunct, or feeder service, to the inter-continental submarine cable system.

At points further East, Singapore and Hongkong fitted well with this concept, each being a compact unit. Regarding

Hongkong, though, there was Great Northern, with whom Pender's telegraph company shared an office, to be considered. Writing to the Governor, Pender's agent suggested introducing an internal telephone service, indicating an intention of combining with Great Northern in the matter.

Sir John Pope Hennessy, once more demonstrating his indifference to mechanical ingenuities, read the letter cursorily and forgot about it. This was in July 1881.

On an upper floor in Queen's Road Central a surveyor named R. G. Alford had premises. Alford's real interest was introducing and selling machinery which was new to the Far East. He was agent in China for the Beaumont Compressed Air Company, for example, and several similar companies, none of them of major importance, yet each in its way making a contribution.

He ordered a consignment of Bell telephones, rigged up a demonstration line to a friend's premises further down the street, and on 28 October that year advertised in the newspapers, inviting inspection of 'these useful and simple instruments', which would be provided and kept in order — and he gave rates for a half-mile and a mile. Wires would connect all subscribers with each other through a central station.

More than three years had gone by since J. M. da Silva's demonstration. There were still no telephones in Hongkong. The value of a line connecting an office with its wharves or godowns was appreciated. As to anything more complicated than this, however, as the editor of the *Hongkong Telegraph* put it: 'There may possibly be some sense or utility in the proposed telephone communication through a central station, but as we are unable to see it, we prefer to remain sceptical for the present.' The owner-editor of this paper, Robert Fraser-Smith, was one of the most forward-looking and dynamic journalists Hongkong has ever had. Those words, coming from him, nicely express the square one from which the telephone exchange started.

Four days before Alford's announcement, Suenson, the Great Northern manager, had written to the Governor, saying it was proposed to set up a telephone exchange in the colony, and seeking his approval. Suenson was awaiting a reply.

His reaction, on reading the papers of 28 October, can be imagined. Before noon he had composed a letter to Alford, and sent it to his office by peon. Suenson informed Alford that

Great Northern were offering Bell telephones for sale, and that the rights which Professor Bell had acquired or was entitled to acquire in Hongkong had been transferred to the Oriental Telephone Company (Pender's London innovation), now amalgamated with Great Northern so far as telephones in Hongkong and China were concerned. (Pender had decided on this limited amalgamation.)

Alford received this. After lunch he composed a letter to the Colonial Secretary, delivered up the hill that afternoon or next morning. Quoting extracts from Suenson's letter, Alford enquired whether the Letters Patent referred to therein had been 'issued in the Colony', and had been filed and registered according to the requirements of Ordinance 14 of 1862. Alford knew his law.

Ten years before, when the telegraph was to be brought to Hongkong, John Pender and those concerned had taken care to ensure that all the various patented inventions giving complete legal security to the cable telegraph had been duly filed and registered under the colony's laws. In the case of the telephone, however, Pender had either forgotten, or else some element in his usually excellent organization had failed to function. Dr Alexander Graham Bell's invention of the telephone was not filed in Hongkong. Anyone could use it, as Alford knew.

Alford's advertisements continued to appear. Then, after ten days, on 7 November, the Colonial Secretary replied to him. Nothing relative to Bell's telephone had been filed, registered, or gazetted. Alford went immediately in person to Robert Fraser-Smith, asking him to publish the complete exchange of letters. The editor of the *Hongkong Telegraph*, one of whose aims was to expose what he considered to be the inordinate degree of sharp practice by European firms in Hongkong — the British were the worst offenders in his opinion — decided to publish the letters.

However, on the same day as the Colonial Secretary had written to Alford, he had also written to Suenson. While Alford was busying himself with the letter he had received, Suenson was dealing with his.

'As far as the rights of the Crown are concerned,' the Colonial Secretary had written, 'the Governor is prepared to grant you permission to establish a Telephone Exchange, but I am to add that His Excellency reserves the right of revoking the permission at any time.' (This was pure Pope Hennessy.)

'You will understand that this permission does not confer on your company any monopoly, but that all other applications to establish Telephone Exchanges will be considered on their own merits.'

Suenson replied immediately, knowing quite enough about the Governor. In a letter dated the same day as the Colonial Secretary's to him, he conveyed his thanks, saying that his company, 'on the virtue of His Excellency's permission', would now invite subscriptions. This letter would have been read in the Secretariat next morning. At 4.30 p.m. that day the *Hongkong Telegraph*, an evening paper, came out with the Alford exchange of letters, showing that Suenson had no rights in Bell telephones.

Suenson handled the embarrassment smoothly. In an admittedly rather lame letter he replied in the *Hongkong Telegraph*, explaining that at the time of the (limited) amalgamation of Great Northern and Oriental Telephone he had received 'letters patent' for Bell telephones in England, India, Australia, China, etc. He was unaware of the Ordinance of 1862, he stated, but was sending for papers from 'home' and would register the patent shortly.

Two days later, on 11 November 1881, he announced the Great Northern Telegraph Company's formation of a telephone exchange, with full details (including the Pender amalgamation), and invited subscriptions from the public. Permission, the statement read, had been given by the Government.

Alford advertisements continued to appear for a few more days, then stopped. There was a rumour that he had been ordered by the Government to take his lines down. He may have, but he had not given up.

Suenson, stressing to the Colonial Secretary the usefulness of telephonic contact with police and fire stations, had offered to link these and other government buildings to the central exchange on specially favourable terms. The Government, while conceding the advantage of this, wished to buy the telephone instruments, while Suenson was prepared only to hire them out. Alford was prepared to sell.

Moreover, with Great Northern's announcement of the coming of the telephone, no new company was formed, no shares were offered. The telephone was to be a blessing conferred upon Hongkong from afar — actually by John Pender who, as mentioned before, thought imperially. The blessing would be paid for by the grateful colonists, the profits remitted to the imperial metropolis. Alford preferred the idea of a local public company for telephones — possibly with himself as managing director.

As Suenson — and Alford — saw it, the hard core of government support for a telephone service lay with the police and fire services, both of them related by the police telegraph. In the police force, Vandeleur Creagh's departure being now assured, tension had eased. Deane was in favour of putting the telephone to the test. He consulted João da Silva, under whose supervision the police telegraph had extended considerably, including to Kowloon by submarine cable.

Silva advised that the existing telegraph wires would serve equally well for telephones. On this basis Suenson demonstrated the telephone between Central and West Point police stations. Deane was impressed. After only brief instruction, even constables could use it.

Alford, worried by this development, invited João da Silva, who he knew to be the key to the government situation, to come to his premises and inspect his telephones which, he impressed upon his visitor, were those used by the English Post Office.

Silva expressed approval, but though he did not say so, he did not care for Alford's prices, and cared still less for Alford's commercial talk. Silva, as a technician, had more confidence in Suenson, with his more technical approach to the subject. They could hire his telephones to get themselves started, then when the system enlarged, buy from London. Deane, on the contrary, favoured buying from Alford.

Even the Governor was becoming interested in telephones by this time. He inclined to favour Suenson. He was thus taken aback somewhat when Deane said he was not in favour of 'a foreign company' controlling government wires.

Sir John reflected. Was there not another company somewhere, a British company? On his instructions a letter was sent to the company which had written before, enquiring if they would care to make a proposition. Back came an answer reminding His Excellency that in July of the previous year they had informed him of their amalgamation with Suenson's company. It was the letter which Pope Hennessy had glanced at cursorily and forgotten about.

Up came the Suenson proposition again.

'Approved conditionally', the Governor wrote, and forgot to put the date.

Deane was mollified by the assurance that Suenson had no

wish to control government instruments. The government system would be their own, and entirely private. As Suenson explained to them: 'If you wish to communicate with any of our subscribers, you can do it by a wire running to our Head Office.'

It was now February 1882, and Alford was still selling telephones. Suenson had applied for permission from the Governor to start a telephone system. Alford knew his law, and was perfectly aware that there was no necessity for Suenson to have done so. There being nothing about telephones in Hongkong law, anyone could start a system. With Alford providing his customers with this reassuring information on the legal position, small telephone systems came into being — wild telephones, as it were. Some merely connected one building with another across the street. Others expanded into small networks connecting several streets. One, exceptionally daring, put up some poles. No more was heard about Alford's public company; it had become public in its own way.

Suenson had somehow to get rid of Alford, and did so thoroughly. The Oriental Telephone Company's rights in Bell telephones having been filed in Hongkong, in September that year Alford was obliged to declare publicly that he had assigned his telephone business to that company, and undertook to import no more telephones, nor engage in any business opposed to the interests of that company.

But this did not put down the mushrooms. By December there appeared to be two systems which were extending across each other, with ingenious structural arrangements to prevent rival wires touching. It looked for a fleeting moment as if the city of Victoria was to have three telephone systems — or four, if one included the governmental one.

At that point someone in the Government went for a walk, or otherwise woke up. Free enterprise in such a matter as telephones, in so small a place, did not make sense. The Government struck. They could not strike the telephones, there being no law on the subject. They struck instead the technicalities: wires laid without permission (there was a subsection of some obscure law about this), and poles — an obstruction, and technically an encroachment on Crown land. Wires and poles were ordered to be removed, including the poles which Suenson, whose service had not yet started, had set up under the impression that he had permission to do so. Order was restored.

Beside this, the more frivolous side of the affair, went another which was more serious.

Gibb, Livingston & Company, one of the oldest-established firms, founded in Canton long before Hongkong became British, wished to have the managing agency of Great Northern's telephone business.

The managing agency was a typical China-coast business feature, designed for incoming companies new to the coast. The managing agent provided office accommodation and staff, canvassed for the new firm, rented residential premises, engaged the servants, handled correspondence, kept the books, ordered and bought anything required, arranged shipments, all in return for $2\frac{1}{2}$ per cent of the firm's gross earnings. It was an iniquitous system — though another fifty years went by before people realized this — because if the firm was successful, the managing agent earned more and more for doing less and less, till he ended by earning a great deal for doing practically nothing. In its finer forms, one clerk could be working for five companies, each of them paying him a salary, himself receiving the salary of one man, while the managing agent pocketed the salaries of the other four. For newcomers to the coast, though, it was a help.

Suenson was not a newcomer, and had no intention of having a managing agent. Moreover, the managing agent was a distinctively British institution. To have one would be compromising for Suenson in the Danish community.

Gibb Livingston was one of the cabal of senior British commercial houses which between them contrived to control everything of any importance in Hongkong. They never succeeded in doing this in Shanghai; the place was too big. Hongkong was just small enough for them to establish a stranglehold. Any newcomer who sought to pursue a course independent of the cabal would be strangled to the point of obedience.

Pender had a managing agent in Hongkong: Gibb Livingston. Why should not Suenson also have a managing agent, the same one? They were amalgamated, after all, in telephones.

It was estimated that there were 100 potential subscribers for telephones in Hongkong; the company actually started with 15. During 1882 Gibb Livingston undoubtedly engaged in indirectly dissuading people from having telephones, as a means of inducing Suenson to appoint them as managing agents, allowing them to do the canvassing for him — bringing him into place, as it were.

They had unexpected allies. Pender's staff on the China coast had never liked the Danish agreement. In precisely that same year, 1882, their dislike swelled into barely concealed resentment. Reports from the north suggested that the Danes had managed to land their cables at Taku and Woosung, which would improve their service (and cut out theft), giving Great Northern more than ever the advantage over Eastern Extension in the China traffic. Worse, there was a rumour that the Danes were secretly laying a land telegraph in China, disguised as a Chinese undertaking. This telegraph was coming south. Shanghai was already connected. It could not be long before there was a telegraph line down to Canton. This would mean that the Danes were intruding in the so-called neutral area (between Shanghai and Hongkong) which in the agreement had been extended to include land as well as sea. For the Danes to do this was not fair, and in secret too.

Now, as will be appreciated, there was nothing secret about it. The Chinese telegraph, with Danish assistance and supervision, was being laid with the full approval of the Viceroy and Grand Secretary, Li Hung-chang. But China was a country in which news — and foreigners did not receive much — arrived garbled. Added to this, in Hongkong, which one might have thought of as a listening post, the paucity of information about China was more marked than in any other foreign settlement. In this garbled form the news of the Chinese telegraph had reached the Eastern Extension men.

With Gibb Livingston as their chairman's agent, Suenson's disinclination to use that company's services for canvassing the telephone was seen, in the light of Danish 'secret' moves in China, as amounting almost to a threat to their chairman's interests. Doubtless they were discreet in what they said to others, yet an observation that the telephone business would not go far without Gibb Livingston's help — and observations of this kind were made — was enough to serve as a dissuasion in Hongkong, where one took care to be in line with the cabal.

One or two Eastern Extension men had for years been writing letters critical of Great Northern to their colleagues in London, in the private hope of their information reaching the chairman, which some of it did. Pender gave attention to missives of this kind. Great Northern had bought used cable, he was told. Its sheathing was rotting, and would soon wear out completely; this was why there were so many interruptions to the service. Nothing was said about cable theft, the real cause of the interruptions.

As Hongkong's telephone service, under Suenson's arrangements, crawled into existence, the letters became more coloured.

'The Danish are so wicked', one of them wrote. 'Mr Pender should not associate with them. Ask Squier, and he will tell you why the Oriental Telephone Company has succeeded ill.'

Suenson had announced the telephone exchange in November 1881. By December 1882, still with no service, he had canvassed 30 subscribers. In that month the Government carried out its clamp-down on poles and wires. After some argument a few of the telephone poles were allowed to remain. Fifteen of the subscribers could not be reached, however, except by means of poles in places where the Government would not allow them. With the remaining fifteen subscribers the telephone exchange started operating on 5 February 1883.

The most cheerful aspect of the occasion was the welcome given to it by Robert Fraser-Smith in the *Hongkong Telegraph*. He, who had been the first to question the need for a telephone exchange, had since become an enthusiastic advocate and the company's first subscriber. He did not mention this in his leading article that day, but he wrote with the enviable distinction of being Hongkong 1.

There was said to be still some telephone competition from small rivals. If there was, it was negligible, and did not last. The Government, which had been publicly criticized for the inept way it had dealt with the telephone affair, would certainly have suppressed such rivals as were left.

More to the point, there do not seem to have been any more new subscribers, or if there were, they were too few in number to justify continuance of the service. It remained in operation till towards the end of the year, when without public announcement it was suspended indefinitely.

It could never be said that the cabal had anything to do with it. That was the cabal's way. All that was left of telephones in Hongkong was the governmental private system.

Chapter 11

Serious Danger from Canton

SIMULTANEOUS with the telephone affair, in the same year, 1882, a related engagement was taking place in the larger dimension of cables and the telegraph in China, with John Pender — thanks to his own achievements, one might say — now much closer to the China scene, although bodily in London.

Basic to the situation was the satisfying fact that the cable telegraph to China, originally seen as something of a hazard, had become a profitable business. Traffic between Hongkong and Shanghai was heavy enough to warrant an additional cable, and Pender intended to lay it.

Report that China proposed to run a land telegraph southward to Canton alarmed him. Hongkong was only 120 miles away from Canton. A land telegraph to Hongkong would surely follow; the bulk of the cable traffic between Hongkong and Shanghai would be diverted to the land line. Most serious of all, the entire China traffic would go to the Danish and Russian lines.

At all costs, as Pender saw it, he must lay a cable from Hongkong up the Pearl River to Canton, and land the cable there, establishing this as his terminal well before the land line reached that far south. This would ensure that at least a portion of the traffic might be carried by the British system, and not be diverted to the Danish and Russian lines.

Another precaution of almost equal urgency was to establish a similar terminal on the Hongkong border. This would entail laying a cable from Hongkong Island under the harbour to the Kowloon peninsula (added to the colony in 1860), the border being along a parallel which bisected the village of Shamshuipo.

British interests — a favoured Pender theme — were, he reasoned, indisputably at stake. Surely at last there would be some British official help, so long and consistently refused. He proceeded to bombard the Foreign Office and the Colonial

Office with letters of interminable length, sometimes at the rate of three a week. The Colonial Office was almost suffocated by it. To them, moreover, 'British interests' meant all too frequently grasping British capitalists trying to coerce native populations. 'British interests — all he wants is a monopoly', one official observed. 'This talk about British interests is another name for the dividends of Mr Pender's Company', remarked another.

The Foreign Office was more receptive. Sir Thomas Wade, British Minister in Peking, contacted by them, assured Chinese agreement to a cable being landed at Canton. He counselled that since, if the land telegraph came south, the cable to Canton would be useless, Pender's company need not be in any hurry.

John Pender's instinct told him otherwise. Revealing in a rare moment the degree of his personal anxiety, he wrote to the Foreign Office:

'My Company attach so much importance to obtaining landing rights on Chinese territory that they are quite prepared to lay a cable between Hongkong and Canton even if it should afterward be found desirable to take it up.'

Wade, in Peking, dismayed at the prospect of his diplomatic telegrams passing to London on the Russian line, considered that Hongkong was the place of real importance. In the same communication he advised the Foreign Office that Governor Pope Hennessy should be warned in cypher to promise no concession to anyone without instruction.

This warning reached the Governor on 10 February. On the 11th he granted a concession to Suenson to lay a cable across the harbour to Kowloon — for telephones.

Sir John Pope Hennessy, once a Member of Parliament, was a man of liberal views. As former Governor of Labuan, in North Borneo, he had formed a high opinion of the Chinese, and of the material prosperity they brought wherever they settled. As Governor of Hongkong he outraged many local Britons by appointing a Chinese to the Legislative Council. As might be expected, Sir John's mistrust of the cabal was only exceeded by the cabal's mistrust of him. He was at this time, among other things, quietly promoting the formation of a Chinese dock company to break the cabal's stranglehold of the dockyards.

In the matter of telephones it will be recalled how he preferred Suenson, who was not of the cabal, and indeed never considered Alford, whom he suspected of being a lowly affiliate of it. (And Sir John was right; Alford was.)

In granting a cable concession to Suenson — for telephones,

Li Hung-chang, Viceroy of Chihli and a Grand Secretary of the Chinese Empire, photographed in Hongkong en route for a state visit to England. A shrewd and capable innovator, to him China owes its first railways and a modern mercantile marine. With Danish technical assistance he introduced the telegraph in 1879.

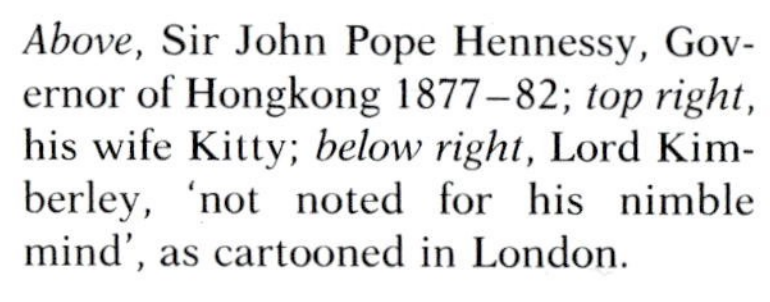

Above, Sir John Pope Hennessy, Governor of Hongkong 1877–82; *top right*, his wife Kitty; *below right*, Lord Kimberley, 'not noted for his nimble mind', as cartooned in London.

Sir John Pender, chairman of the Eastern Telegraph Group, the 'Cable King' who placed 'a girdle round the earth'.

Emma, Lady Pender, *left*, and *above*, son-in-law Sir William des Voeux, Governor of Hongkong 1887–91.

The Penders' London home. They purchased paintings by Turner and other major artists.

The Pender daughters were painted by Millais; Marion, the future Lady des Voeux, is on the right.

an attractive touch — after being warned not to without instructions, the Governor, whose private Chinese sources of information were unusually extensive, had a larger consideration in mind.

This consideration became apparent in Canton three weeks later, when a Canton Hongkong Telegraph Company was announced under sanction of the Viceroy and Government of Kwangtung, to erect a land telegraph from Canton to Hongkong *forthwith*, without waiting for the connecting land link with Chinese points further north, which would not reach as far south as Canton for at least another one or two years.

In London, John Pender's earlier alarm was rearoused. His literary bombardment of the ministries increased to barrage strength. Squier, his Hongkong manager, had cabled him. The new company was a big thing; land lines with extensions to other places; capital $300,000. 'All promoters are natives, Government officials, Merchants and Directors local Chinese Insurance Companies, no European.'

The Governor, he added, seemed evidently inclined to encourage it.

'See Governor Hennessy,' Pender cabled in reply, 'and endeavour to arrange for Chinese Company's line to join our system at Hongkong — we would lay cable to Kowloon — and arrange for this permission.'

It was on this occasion, naturally, that Governor Hennessy indicated that he could grant no cable concessions without awaiting instructions.

Sir John was nearing the end of his governorship. A Briton in Canton spotted the point exactly. The Governor 'wished to sail away, leaving Hongkong's telegraphs safely in Chinese hands'.

He evidently did, and he knew the man who was organizing the Canton telegraph.

'No European', Squier had reported to his chairman. It meant, no Danes. This was the extraordinary part of it. The telegraph from Canton to Hongkong was to be set up entirely by Chinese, without any outside help. Nothing like it had happened before. It was to be a landmark in China's advance into the technical world.

The organizer was Ho Amei, a man with a decent amount of Chinese and English education, and a great deal of polish. He began as a go-getter of the brashest variety, always with good ideas, with the sensibly related aims of general progress and profit for himself. He introduced the first Chinese miners into the New Zealand gold mines in 1866, prospected for minerals in Kwangtung, opened the silver mine on Lantao Island from which Silver Mine Bay takes its name, organized the telegraph, tried to introduce a modern water system in Canton, led the fray in Hongkong for the abolition of the law requiring Chinese to carry a lamp and a pass at night, and steadily mellowed into a notable community leader, an outspoken fighter for civil rights in Hongkong, and a highly respected man. He was the prime mover in the formation of the Chinese Chamber of Commerce, and its first chairman (1897). In these pages we find him midway in his career.

The announcement of the telegraph under sanction of the Viceroy of Kwangtung was Ho Amei's first and very clever move in dealing with the gentry, owners of land along the route. It was a warning to them that if there was any nonsense they were in for trouble. (The Viceroy was to some extent a progressive.) The gentry would demand payment from Ho Amei's company for allowing poles and wires across their land. So might district officials, who would have to be bought. There would be a great deal of haggling, but — with the Viceroy's seal affixed — no outright opposition.

Ho Amei naturally wished his terminal and office to be on Hongkong Island, where the business was. The Canton telegraph would reach Queen's Road Central by means of Suenson's 'telephone' cable which Governor Hennessy had so kindly sanctioned three weeks before. (Pope Hennessy enjoyed Chinese respect to a degree such as Hongkong had never seen.)

Squier's attendance on Sir John, with intent that the terminal of the Canton telegraph be at the colony border, conveyed to His Excellency an instinctive indication of a coming rumpus in Whitehall. He had Ho Amei's request before him. This time he would really have to seek instructions, as he had been warned to do. In a secret cable communication to Lord Kimberley, he referred to the Chinese request that their terminal be on Hongkong Island, saying in so many words: 'I advise allowing it. May I do so?'

John Pender in London, and distantly Sir Thomas Wade in Peking, had between them thoroughly aroused Whitehall. On Lord Kimberley's desk was a filed minute expressing the view that it was vital for the Pender company to have a concession to lay a cable to Kowloon and have their terminal at the colony border, it being of 'the greatest importance that telegraphic communication between Hongkong and the mainland should be in British hands, and that both ends of the cable should be in British territory. A Chinese Company controlling this communication and bringing telegrams into its office in Hongkong independently of all British surveillance might under certain circumstances be a source of serious danger.'

Kimberley agreed. In a secret cypher telegram to Pope Hennessy, he sent his absolute refusal to allow the Chinese telegraph in.

In consequence, a concession was granted to Pender to lay a cable from Hongkong Island to the Kowloon border. From his office in the City came a letter of thanks to the Colonial Office. 'Mr Pender also wishes me to express his gratification at the prompt action of the Government in taking the necessary steps to protect British interests in China.'

If Pender still had not realized the effect which sentiments of this kind had on the Colonial Office, he was perhaps made indistinctly aware of it when in reply came a letter saying that a telegram was being sent to Hongkong on the matter, and they expected it to be sent *free of cost*. Privately they called him 'the monopolist'.

Behind all this was Pender's begrudged agreement with the Danes. When he first learned, late in 1881, that they had acquired cable landing rights in China, he was so incensed that he threatened Great Northern in Copenhagen that if they would not allow him to share the rights he would get the British Minister in Peking to have their concession in China cancelled. This was business bluster, of course, but it was dirty bluster, fouling matters up, as shown very shortly when Suenson, forming his telephone exchange, kept his distance from Pender's Eastern Extension men. These attributed his manner of proceeding to Danish 'secret' moves. The reality was that their chairman had threatened to use the superior power of Britain to have Suenson and his entire outfit thrown out of China; and Suenson knew it.

Pender would have been surprised if he had known the trend of thinking in Whitehall, which was totally opposed to any such *démarche*. At the Colonial Office it was even held that rivalry between the two telegraph companies was desirable, as being of benefit to the Hongkong public.

Pender, whose threat to the Danes had provoked astonishment rather than offers, then made the first moves to obtain independent landing rights in Shanghai for his projected Hongkong–Shanghai cable. When these met with a rebuff, he ascribed it partly to Great Northern opposition. He also partly understood that China, twice defeated by Britain in war, took a cautious view of his advances. Writing to his China manager in Shanghai, he advised him that it was 'not desirable to display personal feelings against Great Northern', and that 'we desire to work pleasantly with the Chinese Government and to be recognized by them as a great European Administration preventing other Companies such as Great Northern obtaining monopolies detrimental to our interests'.

Great Northern were sufficiently disturbed by the Shanghai approach to ask the Danish Minister in London to lodge a complaint regarding Pender's activities, which appeared to them to be hostile. This placed the Foreign Office in the unrealistic position of adjudicators in a matter concerning landing rights on Chinese soil, which had nothing to do with them, on the basis of an agreement between a British and a Danish company to which they themselves were not a party. They wisely declined to adjudicate, finding the embroilments caused by John Pender and his 'British interests' almost as tiresome as the Colonial Office did. All the same, they did support him entirely on the issue of the telegraph link on the Kowloon border.

Ho Amei's wires and poles were meanwhile advancing from Canton in the direction of Hongkong. By June 1882 his company had emerged from the viceregal umbrella as the Wa Hap Telegraph Company, Wa Hap meaning China Union. The name indicated his definite intention of reaching Hongkong Island.

In that month Ho Amei showed his plans to Squier, at the latter's request. The telegraph line, in its approach to the Hongkong region, would go round the Taimoshan massif on the west side, past Castle Peak, and along the coast eastward towards the Kowloon peninsula. Squier explained that his cable from Hongkong Island would land near the Cosmopolitan Docks, which were near the village of Tai Kok Tsui, on the British side of the Kowloon border. He invited Ho Amei to complete his plans, which were indistinct for the last few miles, so as to make the link at the border.

It was almost impossible not to like and be impressed by Ho Amei. He expressed himself well in English, both in conversation and correspondence. He was modern, urbane, experienced. Squier, possibly bemused by talking to so unusual

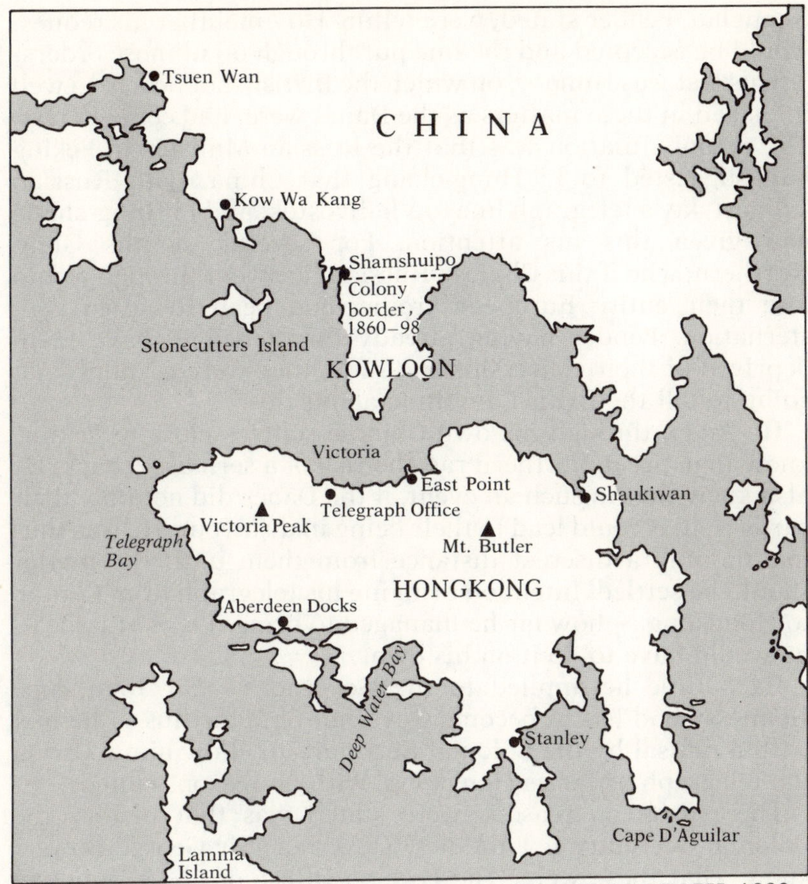

HONGKONG AT THE COMING OF THE CANTON TELEGRAPH, 1883

a Chinese, did not realize till afterwards that his visitor had given no undertaking to make the link.

In some concern, Squier cabled his chairman, saying that Ho Amei was evasive. It indicated to Squier that, although there was no sign of it, Ho Amei must have some kind of tie-in with Great Northern. From afar, Sir Thomas Wade in Peking held the same view, and had already told the Foreign Office.

Pender, heavy-handed once again, sent the Colonial Office what almost amounted to an order to tell the Governor of Hongkong to tell Ho Amei that his request to bring his line from Kowloon to Hongkong was refused. Great Northern in

Shanghai, Pender stated, were telling Ho Amei that his request would be accepted and the line put through on Chinese orders.

This last was rumour, on which the British, not nearly so well informed in these matters as the Danes were, had come to rely. The actual situation was that the Russian Minister in Peking had suggested to Li Hung-chang that China, with Russian support, lay a telegraph line to Vladivostok; and Li Hung-chang had given this his attention. For several months Great Northern, who if this Chinese-Russian line went through would lose their entire European traffic, had been living in consternation. Pender having already threatened to have them deprived of their concession in China, they were certainly not going to tell the British anything about this.

Ho Amei, through his own Chinese sources close to Peking, knew that Great Northern ran the risk of a serious setback. He also knew that in such an event, if the Danes did not play their cards well, it could lead to their being in disfavour. He was thus maintaining a discreet distance from them until the matter should be settled. Intent on bringing his telegraph from Canton to Hongkong — how far he managed to bring it was at issue — he would have to do it on his own.

Each time he applied to the Governor — Sir John Pope Hennessy had left to become Governor of Mauritius — he met with a refusal by the Colonial Secretary to allow him to bring the telegraph as far as Hongkong, with no reasons stated.

The reason no reasons were stated was that neither the Colonial Secretary nor the new Governor, Sir George Bowen, knew what these were. The Foreign Office, on the grounds of 'wider interests involved', simply requested the Colonial Office to refuse each time, giving only rudimentary reasons. The Colonial Office told the Governor of Hongkong to refuse, giving no reasons of any merit.

No one in Hongkong, not even the Governor, knew why the Canton telegraph was being prevented from reaching them.

Ho Amei knew that his Wa Hap company, like all the earlier Chinese telegraph companies, was likely in due course to become part of the Chinese Telegraph Administration. He was thus privately in touch with Director-General Sheng, who was in Shanghai. Sheng, as already mentioned, was the personal nominee of Li Hung-chang, the Grand Secretary in Peking. None of this was known or understood in Hongkong or London, where it was thought that Ho Amei was running an ordinary private company.

Which in fact he was. This was how the Chinese system worked.

At the turn of the year, Squier applied to the Colonial Secretary for landing rights in Hongkong, which he received, for Eastern Extension's cable to Foochow and Shanghai. When a few weeks later this cable was laid, landing rights at both places were refused.

As a courtesy gesture they were allowed to land the cable at Woosung, but were not allowed to use the land line from there to Shanghai, their messages being sent for them at considerable expense. At Foochow the cable rested in a hulk at the point the Danes reached years before, twenty-seven miles from the inland port. Five miles away from the hulk a land line ran to Foochow, but they were not allowed to use it, nor would the Chinese transmit their messages on it. Eastern Extension messages between the hulk and Foochow went by steam launch.

It was quite simple. A Chinese telegraph not being allowed on British territory (Hongkong), the British telegraph was not allowed on the territory of China.

It was less simple to solve, Pender's China manager, Dunn, being under orders not to yield on the Kowloon border issue.

He negotiated in Shanghai with 'the Imperial Chinese authorities' — probably one of Director-General Sheng's subordinates — and reached agreement that the link should be on the Kowloon border. He informed Squier, in Hongkong, of this happy achievement.

Yet when Squier endeavoured to bring matters forward, he found Ho Amei his usual pleasing self, and achieved nothing. On several occasions, Squier reported to his chairman in London, Ho Amei was evasive, a word which irritated Pender.

Ho Amei, almost as if Squier had never spoken to him, made a renewed request to the Colonial Secretary to be allowed to bring the telegraph to Hongkong. He also kept the newspapers informed on the progress of the line, a sensible move which few Chinese of the day would have thought of. Despite being a small line in comparison with the submarine cables, the Canton telegraph was of special importance in Hongkong's day-to-day life. It would be of use to a larger section of the public than any service hitherto.

On 10 May 1883 he announced that the line had reached Tsuen Wan (on the mainland a fair distance from the Kowloon border), and that not being permitted by the authorities to bring the telegraph to Hongkong, this would be the terminal. A steam launch service would carry messages to and forth, a distance of about ten miles.

Predictably, there was an outburst of public annoyance at such an absurdity, and ironically the whole of it was directed at the Colonial Secretary, who was not responsible for the prohibition, did not know the reasons for it, and could not reply.

The 'retrograde policy of the Colonial Secretary' was loudly condemned, Robert Fraser-Smith declaring that Lord Derby, a 'most enlightened statesman', would doubtless have the senseless interdict removed, once it was explained to him.

Lord Derby was now Secretary of State for the Colonies. The Foreign Office having mentioned 'wider interests involved' — which in Whitehall was the same as 'this correspondence is closed' — there was nothing he could do.

Ho Amei, who had been warmly congratulated on his achievement by the Hongkong press and public, decided to bring the line a mile or two further along the coast, to Kow Wa Kang, a hamlet of five or six stone houses above a sandbeach which extended eastward to Laichikok, opposite Stonecutters Island. At Kow Wa Kang he established a well-built matshed terminal.

Shortly before the service opened, he received a report from Canton that five poles had been struck down by lightning. 'Ho Amei', the *Hongkong Telegraph* reported, 'has sent a European engineer with a body of assistants to make the necessary repairs, whereon the line will open for service.'

The European was of course a Dane. Li Hung-chang, in Peking, had quietly dropped the Russian proposition for a line to Vladivostok; Great Northern were out of the woods. A few weeks later they were approaching Governor Bowen, adding their voice to Ho Amei's that the line be put through to Hongkong.

Bowen, an experienced colonial governor, was becoming tired of the Colonial Office's outright refusals. Putting forward the latest request, he said that if he again had to refuse Ho Amei — whom like everyone else he liked — he 'could at least be given some courteous reason for refusal'.

The situation which John Pender had created in Hongkong, as this shows, was socially ridiculous. The Colonial Office in London were fed up with him, in particular with his complete disregard of Hongkong public interests, though they did not feel they could tell Bowen this.

The Wa Hap telegraph service to Canton opened on 9 July 1883, steam launches leaving Hongkong and Kow Wa Kang every two hours, from 8 a.m. to 8 p.m. Charges, at Great Northern rates, were five cents per word to Canton, with — temporarily, it was stated — one cent per word extra to defray the cost of the launch.

John Pender, furious when he learned of it, had the effrontery to write to the Colonial Office demanding that Her Majesty's Government suspend facilities for the Chinese company.

'Yes', mused Lord Derby, 'I suppose the writer means that the steam-launch service should be stopped.' — It was exactly what Pender meant. — 'But we could hardly do that.'

With this piece of Derby statesmanship, the launches between Hongkong and Kow Wa Kang continued to ply.

John Pender had landed himself in a real Chinese muddle, politically insoluble, commercially debilitating, entirely of his own making. He had sought to prevent a Danish monopoly in China. By his high-handedness and misjudgement he had given them one. In the China branch of his worldwide business operations, one more false move and he would lose the game. His ludicrous attempt to stop the steam launch shows all too well that he knew it.

Only one man could extricate him from this. That man was Director-General Sheng Huang-hsün. It is doubtful if Pender ever realized the part that Sheng now played. Pender's men in China sent copious reports of their endeavours and achievements, which doubtless impressed him. In reality, though the Pender men did not know this, they 'achieved' to the extent Sheng wished them to, and in his own time.

Sheng was head of the Chinese Telegraph Administration for nearly forty years, through the fall of the Empire and well into the Republic. He wished China to have an extensive and first-rate internal telegraph system. China's external services would best be handled by foreign concerns, and he wished these to belong to more than one nation, as a precaution of State. He thus in principle welcomed the advent of the British telegraph company.

Britain, however, had twice defeated China in war, and was the most likely nation from which to expect future trouble. It was important therefore that the British should not be permitted to land in their usual arrogant fashion, as Chinese saw it, but enter on a humbler footing, if this could be contrived. Pender, by insisting on monopolizing the line from Hongkong to the Kowloon border (behaving in classic British style, in other words), made it unexpectedly easy for Sheng. By embargoing them at Shanghai and Foochow, he had the British exactly where he wanted them.

Had he been a different man, he could have left them thus, their finances eroding. As it was, having made his political point, Sheng proceeded to the matter which really interested him, which was telegraphy.

A few weeks after the Canton telegraph opened, he privately indicated to Ho Amei his intention of incorporating the Wa Hap telegraph into the Chinese Telegraph Administration, and told him that he could extend his line to the Kowloon border, provided this was done inconspicuously. Sheng then caused a hint to reach Dunn, Pender's China manager, that there existed the possibility of a Shanghai office for Eastern Extension, to be shared with the Chinese Telegraph Administration, and with separate operation.

Having done which, Sheng, in the best mandarin manner, relaxed and let things take their course.

Dunn opened negotiations (as Sheng intended). Obviously, as Dunn saw it, the next move if Eastern Extension were to get anywhere was to offer Ho Amei a shared office in Queen's Road Central, and separate operation. He cabled his chairman in London. Pender, cornered, agreed to this.

Yet when the offer was made in Hongkong, Ho Amei showed surprising indifference. He seemed to be quite satisfied with his existing arrangements. Meanwhile, without moving his terminal at Kow Wa Kang or altering anything, he extended his line to the Kowloon border. This being a wild country area which no one had reason to go near, the British did not notice. Dunn continued to negotiate in Shanghai, making little headway.

Learning of this, Pender saw himself bereft of resources. Humiliating as it was to him, on his orders the offer to Ho Amei was repeated.

This time the clever entrepreneur was almost direct. In addition to the terminal already offered to him in Queen's Road Central, he said, he might also require a terminal, to be constructed for him and to include staff quarters, at the Kowloon border. (His line was through by this time.)

With reluctance, Pender agreed to this too. Ho Amei was informed.

Creating extreme disquiet at Eastern Extension, nothing in Ho Amei's arrangements altered. The border link was not made. Steam launches continued to ply. Unable to give any explanation to their chairman, and knowing his anger and anxiety, Pender's Hongkong subordinates feared for their prospects. (Ho Amei was of course waiting for word from Director-General Sheng.)

The situation being to Sheng's satisfaction, Dunn then attained his shared Shanghai office (as Sheng had intended he should), Dunn cabling to Pender on the successful completion of his exceedingly difficult negotiations. Ho Amei accepted Eastern Extension's offer of two terminals, one in Queen's Road Central, the other at the border. The border link was made. In January 1884 the steam launch service ceased. Canton telegrams could now be sent and received at the Wa Hap Telegraph office in town.

What no one on the British side discovered until it was too late was that Wa Hap's integration into the Chinese Telegraph Administration had been completed. The Imperial Government of China had a telegraph terminal in Queen's Road Central, Hongkong.

It remains to be observed that if Sir John Pope Hennessy had been left to handle it in his own way, this would have been accomplished without fuss.

Chapter 12

A Name and Few Numbers

AMID the outburst of public annoyance which had been directed at the Colonial Secretary the previous year for his 'retrograde policy' in not allowing the Canton telegraph to reach Hongkong, one of the newspapers carried an article about John Pender, whom few in Hongkong had heard of, and none knew was the man responsible for the ridiculous situation which had arisen.

The article was reproduced from an American journal, presenting to its readers Mr John Pender, Member of Parliament, 'the man who has put a girdle round the earth', the Cable King, who was for cable telegraphy what Huskisson had been for railways. 'He has made London the centre of the cable system of the world, and more than any other man, living or dead, has helped to bring every portion of the civilized world into almost instantaneous communication.'

It was a presentation of him which was not in any way exaggerated. Through his various companies he owned one-third of the world's total cable mileage, employing some 1,800 men, of whom about one-third served in his companies' fleet of ten cable ships. Many of these men were to spend the larger part of their working lives on remote islands or in out-of-the-way places with few amenities and only such comfort as they could make for themselves. Most of them went to the companies' training school at Porthcurno in Cornwall, itself a remote place in those times. Arising from this, and from the way in which Pender conducted his companies, the overseas staff became a service, with *esprit de corps* and even an 'old school' feeling. This service spirit was to be of importance in the future; it was one of Pender's signal merits that he created it. He himself was like one of those generals seldom seen yet whom all the men admire.

In the preceding chapters we have observed him in an unbecoming light. This is frequently the fate of men of the West whose interests are sufficiently large to extend to the Far East.

They become entangled in Far Eastern snares, and not knowing what they are entangled by, start to behave at their worst. Pender is one of dozens in this respect.

In fact, if one comes to think of it, it was just as well that people in Hongkong knew so little of John Pender, for seen strictly in relation to Hongkong his record was scarcely designed to win favour. From the start, fifteen years before, his sole interest had been to bring his cable to Shanghai. He brought it to Hongkong solely to make contact with the Chinese Empire, as witness his first telegram to General Whitfield in Hongkong, and in the hope of some day reaching Shanghai. He did everything in his power to prevent the Canton telegraph reaching Hongkong, completely indifferent to Hongkong public interest, as even the Colonial Office realized. Strangest of all, by the accident of his bad relations with Great Northern over landing rights in China, he had wrecked what was in fact his own telephone system in Hongkong. One has to remember that his Oriental Telephone Company was amalgamated with Great Northern in respect of Hongkong telephones.

Let us establish the chronology. The year is 1883. In February the telephone exchange opened; in July the Canton telegraph opened with steam launches; in December the telephone exchange closed; in January 1884 the Canton telegraph was through to Hongkong, and one no longer had to pay one cent per word extra.

Suenson knew, even before the telephone service opened, the secret opposition he was up against; he knew why the service had failed. The catalyst of opposition, in the person of the Gibb Livingston chairman, was the agent of the chairman of Oriental Telephone, with which Suenson was amalgamated. It was an irrational situation. Pender's relations with Great Northern were bad. Suenson's office had lost money on the Hongkong telephone. It could hardly be thought, as Suenson saw it, that Pender had contrived to wreck the telephone service, yet it could be. No one trusted anyone.

What happened to the telephone instruments is not known — a hope had been expressed that the service would resume at some future date. The telephone exchange, at the bottom of Duddell Street, was nominally Oriental Telephone's, actually Gibb Livingston's — the omnivorous managing agent would find other uses for it. Suenson wiped his hands of the business. Nothing more was heard of telephones in Hongkong.

For John Pender, apart from owning the rights in Bell

telephones, the amalgamation with Great Northern was probably nominal. If, when the service failed, there were any financial issues to be settled, it was an embarrassment — anything to do with Great Northern was an embarrassment to him — and he left it to someone else, putting it behind him. In the Pender world, cables were the continents, telephones were islands off the main, telephones in Hongkong were the Antipodes.

Two-and-a-half years later, in Canada in 1886, Pender's son-in-law, Sir William des Voeux — he was Governor of Newfoundland — was offered by the Colonial Office the governorship of Hongkong as his next appointment, which he accepted. Emma Pender and her daughter Marion, Lady des Voeux, were prolific letter-writers — so was Pender when it came to ministries — and this piece of family news (being secret, it could not be sent openly by cable) very quickly reached Arlington Street.

It must have been this which stirred something at the back of Pender's mind, for it was exactly then that he turned his attention to Hongkong telephones.

To savour what he did then, it will be best if we first savour him as he then was. In the pages of the same American journal (Britain was engaged in one of her several Egyptian wars, this one to avenge the death of General Gordon at Khartoum), we can read the following:

> The progress we have made is marvellous [Pender said, speaking of cables]. Time and distance have been annihilated. I was travelling in the Indian country of the United States, in the far West, when the train having pulled up at the small station, my secretary went to the telegraph office for despatches.
>
> I there learned of the magnificent victory of Sir Garnet Wolseley on the very morning on which it had occurred. Was it not almost a miracle? From the Pyramids to the home of the American red man, the lightning harnessed and made to flash thought and intelligence over two great continents and under a great ocean almost quicker than thought itself. I at once cabled my congratulations to Sir Garnet, and I have no doubt he received my despatch before the smoke of the battle cleared off.

It is difficult to think of such a man forming a company for a place with a potential of 100 telephones. Yet with his sense of the comprehensive, the complete, the 'girdle round the earth', Hongkong was an irritating blank space. It was really a question of the company's name — it must not have 'Hongkong' in it — and here one senses the influence of Emma. Marion must be protected from association with a company which would be local and very small. It would be damaging to her — and to William, their son-in-law.

There were precedents for Pender telephone companies having exaggeratedly large names. The Calcutta one, for example, was called the Bengal Telephone Corporation (Bengal was the size of France). A 'China' company would not do. Shanghai, the most important place in China, had had a public telephone service of its own for years. An even larger name, however, might do: a China and Japan Telephone Company. This name, too, had a definite ring to it, more in the tone of one who sends congratulations to generals on battlefields. So thus it was. It was a totally owned subsidiary of his Oriental Telephone Company.

The Antipodean position of Hongkong telephones has been referred to. The reader will not be surprised to learn that the China and Japan Telephone Company reached neither China nor Japan.

It reached Hongkong, however, and with a managing agent: Gibb Livingston. Healthy progress seemed assured. Bearing in mind the earlier confusion of rival telephones, and the British governmental attitude in resisting monopolies, Gibb Livingston saw that a compromise was needed, and reached one. On 1 November 1886 the Government of Hongkong, still under Governor Bowen (he was away), granted to the China and Japan Telephone Company a licence to provide a public service to the city of Victoria. A satisfactory arrangement, to John Pender's way of thinking: with its promise of extension to China and Japan, it would not discompose Marion and William when they arrived next year.

Satisfactory it may have been, except that for three years from the date on which the licence was issued there is no evidence of any other than governmental telephones in Hongkong. These were found also in Kowloon, connected by government cable since 1880 for the police telegraph. When the first cable was replaced by a better one, in 1883, it was for government telephones.

Innumerable publications, governmental, commercial, historical, have stated that Hongkong had a public telephone service from the year 1882. It had a private *government* telephone service from that year. From the closing down of Suenson's service in December 1883, to the beginning of 1890

— a period of six years — there is no evidence of a public telephone system. In the midst of this six-year period stands the 1886 telephone licence. This was lost without trace. The date of it is known; its contents are not known.

By careful scrutiny of the first telephone list (1890) and an examination of what was taking place generally in the preceding years, it is possible to conjecture to a fair extent some of the reasons for the delay which followed the issue of the licence.

Through the Oriental Telephone Company — amalgamated with the former Danish telephone company, sole owner of the present one — the new China and Japan Telephone Company was the direct successor of the earlier one. An understanding of some kind existed with the earlier subscribers that the service would be resumed at some future date. These people were the first who would have to be approached.

Here a problem surfaced which had not existed before. Several of them who had formerly lived in town now lived on the Peak, where there had been pioneer residential development over the past few years. There were nine such persons. Considering that there were only fifteen all told, this was a large number. Four of them were doctors, who would require contact between their Peak residences and their town consulting rooms.

There were chair-paths up the Peak, but no road. One was borne to and fro by chair coolies. The government telephone system extended to the Peak. Mountain Lodge, the Governor's residence, had a telephone, as did the Chief Justice's house, the Peak police post, and one or two others. For the company the foremost consideration was theft of wire. Were they to seek permission to run their lines up the main chair-path, later to be known as Old Peak Road, this would almost certainly meet with police refusal. The police had more than enough trouble already preventing theft of government wires.

A funicular railway, the Peak Tram, which was to be one of the modern wonders of the Far East, was being devised. The only practical way of reaching the Peak residences would be to run the telephone lines up beside it. Work on it had not started. It would not be completed and in service until June 1888 — one-and-a-half years. For the Peak subscribers there would be nothing to do but wait. All of them were prominent men, however. To canvas other subscribers and leave them waiting was inadvisable in a small place like Hongkong. No public

The first homes on the Peak, seen here at 1,700 feet above sea level. *Below*, how you got up there; a studio photograph of the time. A third bearer usually served as a relay.

Mountain Lodge, *above*, the Governor's Peak summer home, was flattened in a storm; its successor had the Peak's first telephone. *Below left*, Observatory and signal station at Kowloon Point, about 1900. *Below right*, the steam-driven Peak Tram, opened in 1888. Telephone lines running beside it connected Peak homes, hospital, and hotel (1890).

announcement had been made concerning the telephone licence. No one knew anything about it. It would be wiser, and not in the least damaging commercially, to set the entire thing aside for the time being, and wait for the Peak Tram.

There was another reason for putting things off. Hongkong was experiencing a boom, amid wild speculation in what proved to be extravagant fantasies, Australian gold (there was very little), Malayan gold (there was none), Borneo plantations (there was no labour force), and so on. It was the first boom in Hongkong's history, and just about the maddest. It was not a time for canvassing anything non-speculative, such as a telephone.

Actually, nothing seems to have been done till after the Peak Tram opened. The aim then was obviously the old one of 100 subscribers. They managed to get about 60. Installation was proceeding when the bottom fell out of the boom in an almighty crash. Several subscribers disappeared — aboard the first ship before landing in gaol — being disconnected before they were connected, as it were, leaving missing numbers in the first list. As the completing touch, the Gibb Livingston chairman, Bendyshe Layton, caught owing thousands of dollars in all directions, absconded — well, 'went on leave' and was not seen again for several years.

In an atmosphere of unsettled dust, China and Japan Telephone seems to have started around 1 January 1890. There were 50 subscribers — it could have been worse. The story of the telephone in Hongkong being a subject which those concerned preferred not to dwell on, congratulations were muted. Robert Fraser-Smith, once more Hongkong 1, recalling the enthusiastic welcome he had given to the first telephone company, kept quiet about the new one, possibly wondering how long it would be before it too closed down. Many companies were closing down; it was the beginning of a long slump.

At the end of the first month the telephone list — numbers, names, addresses, in that order — was printed in the newspapers. This, the telephone directory, became a monthly feature.

At 80 Mexican silver dollars a year (Hongkong had no currency of its own) the telephone service was expensive — company chairman level, one might say. The telephone list did not enlarge much.

Sir William and Lady des Voeux had Government House

connected and listed — Hongkong 44. They already had telephones there, of course. Apart from them and Central Police Station, no government offices or personnel subscribed to the public service. There is even a hint that Hongkong 44 was a gesture to the old man. Mountain Lodge, where a public line could have been socially useful — to call the doctors, for example — was not connected.

PART 3
OUT OF IT ALL

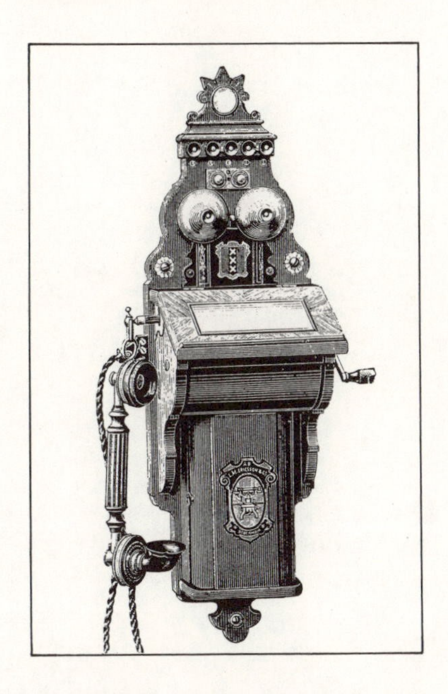

Aeroplanes may not fly over us, because we are a fort. We may not have wireless because we may hear something not intended for us. That Hongkong has never seen a wireless telephone makes us feel very dead. It is not a morbid craving for new sensation that prompts us — it is just a very natural resentment at being forever out of everything.

South China Morning Post
20 June 1922

Previous page: Wall-mounted telephones in brass and mahogany manufactured by Ericsson were a familiar sight in office and home before the automatic exchange. This model dates from around 1900.

Chapter 13

Kowloon More Far

THE telephone was not the only expensive service. Cable charges were monstrous, to use the Colonial Office's word for it. They were the highest charges on earth.

In 1887 Sir John Pender, knighted that year — the year of Queen Victoria's Golden Jubilee — and mellowed in regard to his difficult business in China, mended his relations with Great Northern. Between them, on the basis of a shared purse agreement, their Shanghai managers negotiated with the Chinese Telegraph Administration, obtaining a highly-priced monopoly of communications with China. 'Abnormally high,' the Postmaster-General in London called the rate, 'higher than that charged in any other country in the world.'

Hongkong was included in these arrangements. A telegram from Hongkong to London, at 5s.6d. a word, was not merely per word the most expensive telegram on earth: it was the most expensive to a startling degree. The distance was 9,800 miles. A telegram to London from South Australia (12,000 miles) cost 4s.9d. a word — ninepence less over a longer distance; from San Francisco (6,000 miles) 1s.6d., a much cheaper rate; from Malta (2,280 miles) fourpence a word. At the Malta rate a Hongkong telegram should have cost about 1s.6d. a word, instead of the monstrous 5s.6d. demanded.

Admittedly, over the longer distances cable maintenance and repair costs were higher. Yet against this there was the suspicious fact that Great Northern, with fewer costs on their Russian route, were charging the same rates as Eastern Extension. The entire business looked rigged, and it undoubtedly was.

Over the ensuing years, on various occasions the chambers of commerce of the Treaty Ports, bringing Hongkong in with them, and on one occasion even Calcutta and Bombay as well, petitioned the British Government for assistance in having the cable rates lowered. All to no effect. Questions were asked in Parliament; ministers of the Crown gave bland replies. Perhaps this would have been different if China had not been so far away. And yet perhaps not; for in a subtle way which few people consciously noticed, cables were becoming Empire, part of the nation's greatness, to be upheld, not criticized in small detail. The significance of this change only became apparent after the coming of wireless, as will be seen in a later chapter.

Among Eastern Extension's strong points in Hongkong were its relations with the Government, which were excellent, resting on a basis of special service and assistance provided voluntarily, without the need for negotiation on the Government's part. All government messages were transmitted at half-price. Assistance was given in government cable-laying. All meteorological telegrams were sent free of charge. This evidence of commitment to Hongkong was appreciated by those few who knew of it.

Pender had always insisted on his Hongkong manager dealing with the Governor directly. With his son-in-law as Governor (1887–91) such dealings acquired a naturalness which subsequent managers were careful to maintain with Des Voeux's successors. The cable company, unbeknown to the Hongkong public, had quite a special place in the scheme of things.

The China and Japan Telephone Company was another story. From the outset, nothing about it had been quite right; and this, in commerce as in nature, is important. If seeds are not planted correctly, they will not grow properly. The same is generally true of commercial enterprises.

To start with, Pender's foundation of the company was not a positive act of will; it was the rectification of an omission on his part. Designed to provide telephones for Hongkong, for unexpected family reasons it could not be called the Hongkong Telephone Company; instead it had a ridiculous name, foundationless, a pretence. Next, it found itself with the residue of the earlier company which had failed; in consequence, duly licensed, it proved to be a damp squib. Which, in commerce, is a version of what to expect when seeds are not properly planted.

When at last it began to function, it did so under a serious disadvantage: Gibb Livingston, the managing agents, had no idea how to canvas a Chinese population. The real wealth of Hongkong was Chinese. Ninety per cent of the colony's revenue was derived from Chinese sources. Unless Chinese took the telephone, the company would not go far.

As was observed in Chapter 7, Hongkong Chinese took to the cable service with a will. The telephone presented a more subtle problem. Cables were sent at the cable office; cable replies were delivered by messenger at the office door; either way, there was no intrusion on private property. A telephone, by contrast, involved intrusion. An unknown wiring man would have to come into the office to install it, as well as to repair it in case of need. Chinese mistrusted this proceeding; persuasion would be required if they were to disregard their prejudices. Needless to say, the idea of a telephone in a private house was unthinkable, and with one or two rare exceptions remained so for another thirty years.

In sum, canvassing the telephone needed a Chinese touch. Gibb Livingston's degree of achievement in this field was instanced by there being only two Chinese among the first fifty subscribers, both of them waterfront ship chandlers. China and Japan Telephone's prospects were gloomy.

Stuart Harrison, the first manager, knew even less about the Chinese problem, upon which everything hinged, than Gibb Livingston did. His activity was confined to the very small European community. Even there, response was abysmal. He was manager for fifteen years. For the first ten of these, expansion was at the rate of two new subscribers every three months. It took five years for the total number to rise from fifty to one hundred.

However far away a subscriber might be, Harrison would connect him, on the theory that if one person in an outlying area had a telephone, others would desire to have one. The trouble was, they didn't. After five years he had lines to Pokfulam, Deep Water Bay, and Quarry Bay, as well as the Peak. In the more remote places there was often only one telephone line. The wire mileage was out of all proportion to the number of subscribers.

Then an unforeseeable and totally unnecessary setback occurred. One of their lone lines led to the soap factory at Shaukiwan — seven miles for one subscriber. The owner had another soap factory in Kowloon, and in 1896 asked Harrison if there was a chance of having a telephone line there as well.

Much of the port — docks, wharves, godowns — was on Kowloon peninsula, where in addition there was an attractive garden suburb inhabited by Portuguese and other resident communities, leading more settled lives than did the expatriate Europeans with their endless succession of leaves, postings, and replacements. Kowloon might prove to be a first step to bringing the telephone into the homes of the resident population.

Government permission would be required to lay a cross-harbour cable. In the ordinary way, this would involve a direct request to the Secretariat. Harrison, however, was under orders that all policy matters between his company and the colonial authorities must be conducted in London. Here again one senses the influence of Emma Pender protecting her daughter Marion. The link between Marion des Voeux and the telephone company had to be kept out of sight. The Des Voeux governorship had ended five years before, yet the injunction to deal solely through London remained.

Possibly, on this occasion, Harrison did not word his cable to London clearly. However it came about, China and Japan Telephone showed the first sign of the inadequacy which was to become their permanent feature, and which thirty years later led to their being virtually thrown out of Hongkong. When they wrote to the Colonial Office, instead of asking to lay a cable, they enquired whether Kowloon was an integral part of the city of Victoria.

The 1886 licence was for the city of Victoria, beyond the boundaries of which — for instance, on the Peak — the telephone company had been functioning since the first day of its existence. Stuart Harrison had treated the licence as meaning 'and suburbs'; the Government had raised no objection; there was no reason why Kowloon should not be treated as yet another suburb, which, as just explained, was precisely what it was.

It was not, however, 'an integral part of the city of Victoria', from which it was separated by more than a mile of sea — as the Colonial Office informed them. The inadequacy of China and Japan Telephone, demonstrated in this instance by getting hold of the wrong end of the stick, rebounded at first solely upon themselves. By asking the wrong question, they had placed themselves in the position of having to apply for a separate Kowloon licence.

It went further than this, however, for at that moment Sir John Pender died.

He was eighty-one. He had revolutionized the communications of the world. He had also created a commercial empire, diverse and complicated — the ships, the cable stations, the cable-making plant, the telegraph offices, the telephone exchanges, the personnel for all of it — all held together by one man, himself.

A succession on such a scale as this inescapably causes anxiety to employees. More than a few, even in the group's London offices, wondered whether this might not be the beginning of a disintegration, or a separation into smaller elements with different loyalties.

Anxiety, like a rebuke, is augmented by distance. In Hongkong, 9,800 miles away, the Eastern Extension men felt insecure. To maintain cohesion, to keep their identity, they closed ranks. When they learned of Harrison's intention to connect Kowloon, they protested in a form of self-defence. An agreement existed, they maintained, between their company and the Oriental Telephone Company, parent of China and Japan Telephone, that the latter would not run a service between one city and another.

Although Harrison did not know it, this agreement was a devolved remnant of the original telephone arrangements for India, in which each major city would have its own telephone system, communication between one city and another being by telegraph. To apply it to Hongkong and Kowloon was a complete misinterpretation, not even sustainable in India, where Calcutta and Howrah, two cities separated by water, had the same telephone system.

Harrison, judging by his actions, saw two points with clarity. If he went ahead with his application against opposition from Eastern Extension, who carried weight with the Government, then the subject of the telephone licence was almost certain to come up, which would embarrass the Secretariat quite as much as it would his own company. More saliently, it could be argued that this dispute between two companies which operated internationally was not of specific concern to the Hongkong Government.

He thus turned to the Chamber of Commerce, asking them if they would arbitrate in the matter. At this point the dispute became public. It had also become thoroughly unpleasant, evidently due to personalities. Eastern Extension presented an adamant case to the Chamber. It was clear that on no account were they going to give in.

The Chamber of Commerce, confronted with a case which it would have taken months of legal research to give an opinion on, took refuge behind one of their rules which allowed arbitration solely between members. Neither company was a member.

For Harrison there was no other forum of appeal. In London, China and Japan Telephone were in a state of suspense caused by Sir John Pender's death. In Hongkong, the same event had caused two of his companies, formerly on the friendliest terms, to become opponents, and this against all logical reason.

Kowloon was not connected with the telephone service for another ten years.

Chapter 14

Circuits and Listeners

By treaty with China, 355 square miles of land were added to the colony of Hongkong in 1898 on a 99-year lease. The New Territories, as the leased land came to be called, consisted of a sizeable piece of mountainous, sparsely populated mainland behind Kowloon, and numerous islands around Hongkong.

There was only one town in the New Territories. This was on the little island of Cheung Chau, seven miles south-west of Hongkong. Notorious for sixty years as a pirate transit base, it was the first place to have a police post when the British took over; at an early date a telephone cable was laid to it. Police stations were established at well-chosen strategic points on the mainland; near each of these in the course of time small market towns developed. The mainland police stations were connected by telephone with Kowloon and with each other. By 1901 the Government ran 150 miles of telephone lines.

When it was found that lightning was doing an excessive amount of damage to the poles — in ricefield areas of wet cultivation, side-flashing wrecked the timber — the entire mainland system was reconstructed with iron poles. In sum, the Government had displayed a good deal of energy in a short space of time, and was justifiably pleased with the results.

While this was going on, China and Japan Telephone reached 150 subscribers, and altered the order of things in their monthly lists in the newspapers so that they read name, address, number — a faint sign of progress. With such a slow rate of growth, though, no technical improvements had been made in the system, which was at the very least twelve years out of date. This, and a good deal more concerning the company's shortcomings, came to light in the year 1901.

A tramway system for Hongkong Island had been proposed. The promoters, a London company, had a tie-in with an outstanding Hongkong entrepreneur, Robert Shewan, and themselves differed impressively from Pender's Oriental Telephone Company and its colonial offshoots. The trams were to be the best-designed, the most comfortable, and the cheapest trams in the world.

The company's technical consultants arrived in Hongkong in January 1901. They were received warmly by Sir Henry Blake, the Governor, who saw trams in terms of cheap transport for working people, and was keen to see the scheme go through. A Tramways Ordinance had already been drafted, with a clause safeguarding concerns with electric systems which might be interfered with by the tramway circuits.

Eastern Extension, the Hongkong Electric Company, and others, immediately had themselves included under this clause. When it came to telephones, the consultants made a sharp distinction. While they were prepared to accept the protection clause in regard to government telephone lines, China and Japan Telephone could not be included unless they modernized their system, and in particular did away with earth wiring. As they explained, 'If we agreed, it would permit the Telephone Company to add *ad infinitum* to their present obsolete system of using the earth return, and calling on us to provide a metallic circuit every time there was interference to their system.'

This, as gradually became apparent, was just what the London directors of China and Japan Telephone had in mind. There would be no necessity for them to make technical improvements. It would be done for them by the tramways.

When Stuart Harrison asked for his company to be included under the clause, this was refused. The consultants offered him £1,500 against the cost of putting in metallic circuits — about one-quarter of the actual cost. This offer Harrison referred to his directors in London.

There followed eleven months of guarded replies which neither accepted nor refused, interspersed with long silences. Everything was ready for the introduction of the tramways. Only this one company stood in the way. The consultants raised their offer to £3,000, pointing out as they did so that in Britain no tramway company would be required to pay them anything at all — which was correct — and giving them a deadline for acceptance. This they received in silence.

On the deadline date, instead of writing to the tramway consultants, they wrote to the Colonial Office. The chairman and secretary of the China and Japan Telephone Company asked for an interview with the Under-Secretary of State to show him their correspondence with the builders of the trams. The chairman, Sir Auckland Colvin, 'is ready to furnish the

Colonial Office with information, and call there at any time convenient to them'.

This kind of approach was enough to put anyone off, and none more surely than Whitehall. Sir Auckland Colvin, who had caused the delay of the introduction of the tramways into Hongkong by nearly one year, was refused an interview with the Under-Secretary of State, and was given one week in which to settle his affairs with the tramways company. At the same time, and within twelve hours, the tramways' request for a 50-year concession was submitted to the Secretary of State, Joseph Chamberlain, who agreed to it at once (23 February 1902), leaving the telephone company to get on with it.

This was a moment when people with long memories in Hongkong wondered whether for a second time a telephone company was about to close down. By what must have seemed to Stuart Harrison something like a miracle, just then the rate of applications for telephone service rose, *and the applicants were Chinese.* By the end of 1902 the company seems to have had some 380 subscribers on their books (not all of them connected), and by 1905, when Harrison at last succeeded in connecting Kowloon, the number had risen to 500.

The oddest feature of this development — it was the hinge of survival — was that neither Harrison nor any of the Europeans concerned knew to their dying day what caused it. In the company's later reports it appeared as, 'In that year Chinese started taking telephone service', which in the colonial ambience indicated that they were praiseworthy Chinese issuing out of their retarded state. Equally in the colonial ambience, Britons involved could not help sneakingly praising themselves for this happy evidence of progress. What actually happened was this.

China and Japan Telephone, a London company, and the Hongkong Electric Company, a local concern, opened for service in the same year, 1890, in January and December respectively. Both had the same managing agent, Gibb Livingston. Both were faced with the same Chinese problem of intrusion on private property. Both were victims of Gibb Livingston's incompetence at canvassing (the two waterfront ship chandlers). Both endured twelve years of almost non-growth. The electric company during those years reckoned their rate of growth by the number of light bulbs.

Leung Yan-po, the Gibb Livingston compradore — a compradore was an independent business man, not an employee — had the ledgers of these two unprofitable companies cluttering up his office, and as he related many years later, grew tired of them. Without saying anything to the Gibb Livingston people, he did a deal with the wiring contractors, enabling them to take a cut on each new line they connected. It worked like magic, and the Europeans never found out.

There was a difference, however, in the manner in which the two companies then developed, and a difference in Leung Yan-po's attitude to them. Hongkong Electric developed on a much larger and healthier scale than the telephone company, and every few years, when they could afford to, lowered the electric rate, causing further bursts of new applications, until by 1923 Hongkong Island had the cheapest electricity in the world, a distinction shared with Shanghai. The telephone service remained expensive, and after the initial breakthrough of 1902 it grew only slowly.

One of the reasons for this — more significant than the expense — seems to have been that Leung Yan-po was not interested in it (even in negative ways the compradores wielded influence, and Leung was one of the most influential of his day). A new manager took over China and Japan Telephone in 1905, from which time a marked feature of the company was secretiveness in the conduct of its affairs, a factor which contributed to its ultimate demise. The suggestion is that Leung Yan-po was excluded from the new manager's confidence. Bearing in mind that the Hongkong Electric manager's office was two doors away on the same floor as that of Leung Yan-po, who by the way held a large block of shares in the electric company, the disparate development of the two companies becomes clearer.

The telephone directory was by this time a printed booklet, with advertisements on every other page. The service cost more than double what it cost in Shanghai, but this was the way of anything which originated in the Pender empire, where the foremost consideration was dividends. Rates were invariably pitched as high as it was thought the public could bear.

The Government was aware of this aspect; the tramway affair had opened their eyes to a good deal. In 1905, when Kowloon was connected, they insisted that the former licence be terminated, and replaced by a new one — for 25 years — of which a feature was that the subscription rate could not be altered (in other words raised, because they would never have lowered it) without government permission. On this basis of warning to the company, telephone business quietly grew, till by the year 1922 there were 4,000 subscribers, the population of Hongkong having risen by that time to 650,000. Three-quarters of the subscribers were Chinese.

Stephen Grove, in his unpublished account of Hongkong's telephone companies, gave a perceptive view of the times. Of the British, he wrote:

It is very likely that the community must take part of the blame for the slow progress. Hongkong existed purely as an entrepôt port, and all business life was geared to the arrival and departure of ships and the length of time it took for mail to and from the United Kingdom. There was thus no great sense of urgency in business, so that it was unlikely that businessmen felt any particular need for quick means of communication, a chit coolie being sufficiently fast and reliable, and far cheaper than a telephone, in addition to being possibly more reliable and less frustrating. The same circumstances engendered an air of sleepiness in the Colony which also possibly led to the ladies not displaying any particular interest in having telephones in their residences, being content to rely upon the services of a multiplicity of servants. These circumstances prevailed in some degree right up to the outbreak of the Second World War. In contrast, Shanghai was always a busy and bustling city, and in consequence it must have been far easier to sell telephone service there.

Complaints about the government telephone system — complaints from government officials — mounted to a point when in 1914 it was decided in the Secretariat that the system was unsatisfactory and needed restructuring. Complaints centred on 'hearing over the lines, due to induction or contact with the aerial wires, or negligence of operators, or both'.

There were now 218 miles of aerial wires, of which 120 miles were in 'the most congested part of the city'. The system connected 129 offices and several official residences, from ten exchanges. In addition there was the cable mileage to Kowloon, Cheung Chau, and the lighthouses — Gap Rock (40 miles) and the nearer ones on Waglan (7 miles) and Green Island. London technical advice being needed, the public works chief wrote a description of the system, simple and excellent. Someone at the Colonial Office shuddered — 'This appears to me to be a highly technical matter' — and sent it away at once to the Crown Agents, who sent it to Preece, Cardew & Rider, the foremost electrical engineering consultants of the day.

There was a pause lasting several months. Then, from Preece Cardew, came a faintly bewildered reply, saying they did not entirely understand the description. Surely there must be some kind of central exchange, they wrote. How did one exchange contact another?

Well, of course, there was no central exchange, and there never had been. If you, at the Secretariat, wished to call the Stores Department, you lifted the ear-piece of your telephone off its hook, and asked the Secretariat operator for Stores. He called the Treasury, or perhaps the Buildings and Lands Office, where the operator might connect the Wanchai post office, whose operator connected the Stores. It was very simple when you understood it.

There was the drawback that, while you were on the line, the Attorney-General, perhaps by way of the Fire Brigade, might ring you. This being an incoming call to which the Secretariat operator gave a conditioned response, he was almost certain to connect you, interrupting your call to Stores.

There was another slight drawback. The operator of each exchange was a clerk with other duties to attend to; the exchange was on his desk. If he was called away for any reason, the exchange was left unattended. On his return, assuming that everyone must have stopped talking, he would disconnect all the lines. Senior officials did not mind about this; government offices were placid places. What they did mind about was that in the clerk-operators' placid hours they *listened* to the conversations. This had to be stopped.

Upon further consideration, they began to see it more clearly. Interference with aerial wires (caused by overhanging structures from houses, or by poles having to be removed to clear construction sites) could be remedied by going underground. A central exchange was a good idea; it would remedy the rest of the problem. Room 16 on the top floor of the Post Office building was vacant. The exchange could go there, all lines being connected with it. At present there were eleven operators for ten exchanges. With a central exchange and a competent supervisor the number of operators could be reduced to four — three by day, one by night. They would be kept so busy that they would have no time to listen to the conversations.

It was not that easy, however. Britain was in the throes of the Great War of 1914–18. So many factories had been made over to war production that it was all but impossible to obtain mechanical equipment for civilian use, especially in faraway colonies. Preece Cardew reluctantly advised that it meant waiting.

By 1919, when the supply situation showed the first signs of

Sir Paul and Lady Chater, *above*, at their home, Marble Hall (No 1 Conduit Road), which was completed in 1902.

Left, James Taggart and his wife, who was Californian, at the Repulse Bay Hotel, of which he was a director, about 1928.

Exchange Building, later known as Telephone House and then as Lane Crawford House, on Des Voeux Road Central; the 1930 automatic exchange was installed in the top floor.

The General Post Office, on Pedder Street, ran the colony's wireless services, 1921–37, and housed the first broadcasting studio (1928).

easing, the Colonial Secretary, Sir Claud Severn, had a better idea. The automatic exchange had come into being. It was reported that New York, with nearly a million telephones, was going over to an automatic system. Tokyo, with 300,000 telephones, had started going over to automatic in 1915. Would the Crown Agents kindly obtain advice on the feasibility of the Government of Hongkong having an automatic exchange. The Secretariat would shortly be moving into a large new building on Lower Albert Road, where there would be ample space for the proposed central exchange. Consideration should be given to its being automatic, and there could be another automatic exchange for government lines on Kowloon Point, and another for the police in the New Territories.

Grateful early reply.

There was no early reply. When a reply came, more than one year later, it was to the effect that Hongkong's humidity was so high that no automatic exchange was ever likely to function. A central exchange was conceded, but it would be manual.

By the time this dispatch was received, Hongkong Electric had installed a 100-line automatic telephone exchange of their own, made by Siemens. It functioned perfectly.

Chapter 15

A Luncheon in London

EVER since the tramway affair of 1901–2 the Secretariat had had its doubts about the China and Japan Telephone Company, as it made clear in 1905 when it imposed governmental control of the subscription rate in the licence issued that year.

Hongkong was on a silver currency (conforming with China) which was the bane of everyone's lives, owing to silver's continual small fluctuations and occasionally drastic ones. Telephone subscriptions were paid in silver dollars.

In 1907, silver having dropped severely against sterling, the company asked the Government for permission to render accounts to their subscribers in sterling at a deemed rate of 10 silver dollars to the pound, which had been the rate of exchange at the time of the 1905 licence. The Secretariat refused. Everyone in Hongkong was in the same boat currency-wise. There seemed to be no grounds for permitting a special rescue operation for one company, particularly for one which provided an expensive luxury service.

Silver having sunk still lower, the company in 1910 renewed their plea, saying that they were losing money badly, needed to improve the system, and simply did not have the funds. This time the Secretariat relented. Accounts were from then on presented in sterling.

Next came the Great War of 1914–18. It had little or no effect on the Far East — in Shanghai these were in fact boom years — but the value of silver soared. Once again, in 1917, the company pleaded that they were losing money, begging this time to be allowed to go back to silver.

The Secretariat, deciding that they were tired of this shilly-shallying between currencies, asked to see the company's books — something they had not done before. These, when presented, showed that over recent years their profits had been so good that only five months before, in London, they had distributed the company's entire reserve fund as a bonus to shareholders. Needless to say, their request was turned down peremptorily.

Oblivious of the bad impression they had given to the Government — they were thick-skinned to an amazing extent — in 1919 China and Japan Telephone pleaded again, this time in desperate terms, saying that their situation was so serious that they were 'practically unable to carry on, suffering very heavily from the very high rate of exchange'.

The Government had had enough. William Fletcher, who was the Clerk of Councils, and who on occasion stood in as Colonial Secretary when Sir Claud Severn was away, called in the manager. Before him on his desk Fletcher had the company's annual report, recently published in London, which stated that the years since 1905 had been marked by 'great development and prosperity of the Company's business'. Dismissing out of hand the contention that the company was making losses, he added for good measure that the service was not all it might be — it was so bad that it is small wonder there were only 4,000 subscribers — and proposed that 'in order to get rid of this trouble of the fluctuation of exchange' the company should move its headquarters to Hongkong, and work on a silver dollar basis like everyone else.

This, of course, was exactly what the company was determined not to do. With their headquarters in Hongkong, they would be exposed to public scrutiny. For fifteen years they had been raking in profit with an air of despairing loss. They had every intention of continuing to do so. Colonies, in any event, were simple. The people there had enormous respect for London, and for London institutions. Send rather a senior man there — giving them face, as it were — and he would properly sort matters out.

This brought to the Far East the vice-chairman of the Oriental Telephone Company (China and Japan Telephone's parent), by name Parker Ness, a barrister who in everything he did and said conveyed an unassailable sense of his own superiority. Seldom in the history of Hongkong has a visitor so infuriated those segments of the population who had the misfortune to be aware of his existence. His utterances were lordly to a degree. He boomed more loudly than the Jardine noonday gun. Withal he was completely insensitive to his surroundings. He trod on every colonial toe imaginable. It amounted to a gift. As someone memorably commented afterwards, 'Every time he opened his mouth, he put his foot in it.'

In the Secretariat, to which he addressed himself, the effect he produced was worse than anywhere, being more serious. He arrived bringing with him a draft agreement for a 50-year extension of the telephone licence. In this there was to be no governmental control over the rate charged. The agreement, he explained, was on the same lines as one which he had recently concluded in Singapore. He produced an authenticated copy of this agreement.

The Secretariat of those days (February 1920) was run by four officials. Such was the mistrust which the barrister inspired that none of those who saw the contract copy believed it to be genuine. Politely getting rid of him, they communicated with the Straits Settlements, whence after some days word was received that such an agreement had in fact been signed. Strange to relate, this if anything deepened mistrust of the barrister. He was called in and told that, regardless of anything to do with Singapore, there was no possibility of the Hongkong Government withdrawing its control over the rate. Parker Ness withdrew for a week.

He returned with a proposition to raise the rate, in silver dollars, by what amounted to two-thirds, 'in view of increased costs since the end of the war' (this was a theme-song of the day), making no reference to moving the company to Hongkong. His attention was drawn to his balance sheets and the evidently prosperous state of his company. Someone — probably Fletcher — told him that the Government could see no reason for an increase in the rate.

The visit had aroused the public's latent dissatisfaction with the telephone service. An idea was floating about in commercial circles of buying out the company. It would be £9 for a £1 share, Parker Ness intimated, thereby effectively dismissing the matter, and sailed away to England.

As things stood, Hongkong was saddled with this company for another ten years, until 1930 when its licence would expire. In the Secretariat they would have liked to urge matters forward. Ridding a colony of a public utility being not exactly in the ordinary run of government business, they were unsure of themselves. In a stage whisper they let the Chamber of Commerce know how much they would appreciate a local company being formed to buy out the telephone business. This coincided (August 1920) with an American market crash which hit Far Eastern commerce severely. The Chamber intimated that nothing could be done for the moment.

The Secretariat next appointed a public committee to advise on the telephone rate. Its members included Frank Marsh, Hongkong Electric's able manager. Not even he could pierce the secretiveness of the telephone company, which refused to reveal anything concerning its financial workings. After several sterile months the committee recommended a slight increase in the rate, and agreed with Sir Claud Severn's view of the need for the company to have an automatic exchange.

This was an outlay which the China and Japan Telephone Company had no intention of making. Informing the Government that this was a matter of a 'highly technical nature', they said they would be prepared to seek professional advice on it, but only after terms for a revised licence had been agreed. Even then, it would depend on their consultants; they would not be bound to an automatic system. And they declined to agree to the rate proposed by Marsh's committee.

The colony was thus stuck with this entirely inadequate telephone company. The company, however, was stuck with the rate that caused them such a 'loss'. Somehow they had to become unstuck.

In the summer of 1921 Fletcher went on leave. In June he was entertained to luncheon in London by Parker Ness, who was surprisingly forthcoming. Giving £325,000 as the company's valuation — its actual value was about £90,000 — he led the conversation to the appointment of two independent advisers, one for the company, the other for the Hongkong Government, to give an agreed valuation of the company, and an agreed rate.

The two advisers duly conducted their deliberations in Hongkong. The company chose a London consulting engineer; the Government chose Shanghai's telephone manager. The consulting engineer either outwitted the telephone manager, or more possibly caused inducements to be offered to him. Whichever it was, he agreed on a final figure which he knew to be inflated: £288,000 with a much increased rate.

This was not all that resulted from that luncheon in London, however. Parker Ness took Fletcher thoroughly into his confidence. Expatiating on the numerous complexities of telephone companies, their financing and costs, citing examples of the telephone systems in India and elsewhere, he eventually tricked Fletcher, who had no experience of the cut and thrust of commerce, into agreeing that there was a *prima facie* case in Hongkong for doubling the rate.

With this in hand, Parker Ness booked an autumn sailing to

Bombay, mail train to Calcutta, and onward sailing to Singapore. In Calcutta and Singapore he would brandish Hongkong's agreement to higher telephone rates in justification of higher rates there, just as he had earlier brandished Singapore's agreement before Hongkong. Fletcher would have a lot of explaining to do about that luncheon.

The Bengal Telephone Corporation (Calcutta and Howrah) doubled its rates. Singapore, which had been promised a lower rate, was given a higher one. In March 1922 Parker Ness sailed for the second time into Hongkong.

With perfect timing, the two 'independent' advisers presented their report to the Government. Parker Ness was jubilant. The agreed rate would amount to a doubling, while the agreed valuation of the company was three times more than it was worth. He prevailed upon the Secretariat, in advance of the issue of a revised licence, to allow the consultants' findings to be publicly announced, and that in the interests of improving the service, the sooner the new rate was introduced the better, say on 1 July.

The Secretariat, honestly trying to do their best in a matter which was beyond them, saw no objection. The announcement was made.

The press uproar which ensued had no previous parallel in Hongkong's history. The newspapers were deluged with letters of protest, complaints about the shabby service, faulty economics, arrogance — this meant Parker Ness — and the preposterous demand that subscribers should have to pay in advance for improvements promised for the future. It went on for weeks. Entire pages were consumed by it. All concerned caught the blast, including the Government, which was accused of going about things in a hole-in-the-corner way, and not protecting the public interest.

The company was particularly blamed for the poor standard of its operators. A Chinese with sufficient knowledge of English to be a competent operator commanded a price in Hongkong which the company was not prepared to pay. In consequence, they were getting unmannered semi-literates only one degree removed from the underworld.

Operators [one correspondent wrote] repeat the number at such speed (and connect) that there is no possibility of correcting him if he has got it wrong. 'Engaged' is given at the slightest whim; it can take an hour to make a Kowloon call. Orders made by houseboys are totally ignored or deliberately misplaced. Things have got to such a pass in my household that I, with others, dread to use the 'phone, and unless the matter is pressing, do not use it.

Making a call at night was almost impossible because the operators were asleep. If a Hongkong Electric man wanted to make an outside call at night and could get no response, he would dial — on their Siemens exchange — the Duddell Street electric substation engineer's flat, which was across the street from the telephone exchange and on the same level, and ask the engineer to wake the operators up. There they were, sound asleep in front of their switches, lights flashing. Duddell Street being very quiet at night, the engineer avoided shouting. He hurled a lump of coal at them.

It happened that the outburst in Hongkong occurred at the very moment when Calcutta and Singapore suddenly woke up to what had been done to them. The outburst in Calcutta was such that the British and Indian mercantile communities, who despised and detested each other, were linked as if they were old comrades in a forgotten war. The Rotary Club, in a protest meeting, even seems to have invited some Indians. The Maniktolla district of Calcutta told the Bengal Telephone Corporation that it could take its instruments away. Nobody had ever seen anything like it. It, too, went on for weeks, and naturally the Hongkong newspapers chronicled every day of it, to regale Parker Ness at breakfast.

It had no effect whatever. When the Government told him that, in view of the public clamour, there could be no question of them allowing the proposed 1 July increase in the rate, he replied in writing, in smooth and suave words, accusing them of breach of faith. As to the Hongkong public, he informed them that he had warned the Government that unless another licence was issued on the lines he had proposed, the service would simply run down until 1930, when the cost of improving it would be so high that subscribers would face an increased rate far in excess of the one he had proposed, and which the Government had refused to allow.

On this mingled note of warning and threat he departed.

A joint committee of the two chambers of commerce, European and Chinese, had been set up to consider the operation of the telephone. The Secretariat, contrite after the press hammering it had received for its own efforts, spontaneously said it would be guided by the joint committee's decisions. These recommended an extension of licence for the company beyond 1930, provided it undertook to update its system at once and provide an efficient service, submit to control of the rate — and it laid down a rate — and keep its books in an approved form. If the company did not accept these terms, its licence should not be renewed.

The company did not accept. The Colonial Secretary promptly tabled a resolution in the Legislative Council to the effect that unless the company did accept, no further agreement would be made with it. This was carried unanimously in October 1922.

The company made counter-proposals which the committee informed them were unacceptable. Thereupon the company once more resorted to silence. For a year not a word was heard from them.

Not even a colonial government could be expected to put up with this.

Chapter 16

A Law No One Looked at

A year later, in January 1924, we find James Taggart, a hotelier, and Sir Paul Chater, Hongkong's most eminent citizen, promoting the formation of the Hongkong Telephone Company, which was to be a local undertaking. Exactly how this came about is unknown. It is known, however, that while Taggart was taking the lead, Sir Paul Chater, much his senior in years and estate, had invited him to do so.

Chater, who was seventy-seven, was a Member of the Governor's Executive Council. The likeliest probability is that it was from there — possibly even from the Governor, Sir Reginald Stubbs — that a request was made for his advice on the telephone situation. Alternatively, it could have been spontaneous on Chater's part. There is no clear evidence either way. What is certain is that the most important man in Hongkong now addressed himself to the matter.

When Chater died a few years later, the *South China Morning Post* observed in its obituary of him that 'a biography of Sir Paul Chater would be a history of Hongkong', which was true. An Armenian born in Calcutta, he first came to Hongkong in 1864, aged eighteen, made his fortune as an exchange and bullion broker, and proceeded to identify himself with his new home to such an extent that it would be no exaggeration to say that he thought *for* Hongkong. His ideas and efforts lay behind every major development in the colony's history from then on, including the acquisition of the New Territories, for which he even delineated the boundary. As a Member of Council he advised eleven successive Governors, a fact which perhaps more than any other testifies to the uniqueness of his position. There is another extraordinary fact about him: in his long and successful life there is no trace that he ever made an enemy.

So much for Sir Paul for the moment. How did James Taggart, the hotelier, come into the telephone picture? Who was he, and why had Sir Paul invited him to be his fellow-promoter?

A Lowland Scot of evidently very humble parentage, James Harper Taggart had innate good manners, a feature which is mentioned in every known description of him. Of less than medium height, upright in bearing, and smart to the extent of being called dapper, he had a natural gift for organization and management. He came up in the hotel business the hard way — what many hoteliers would say is the only way. First heard of in the kitchens of a hotel in West Africa, he reached Hongkong as a saloon steward in a Canadian Pacific liner, signed off, and joined the staff of the Hongkong Hotel, of which he became bar manager, and subsequently hotel manager.

The hotel — the colony's best — was part of the commercial dominion of the Kadoorie brothers: Sir Elly Kadoorie in Shanghai, Sir Ellis Kadoorie in Hongkong. In 1923 Taggart was made managing director of the Kadoorie hotel group, Hongkong and Shanghai Hotels, which owned the Wagon-Lits in Peking, the Park in Shanghai, and had just opened the Repulse Bay in Hongkong. They were about to embark on their major hotel enterprise, the Peninsula in Kowloon.

The Peninsula Hotel was the joint brainchild of Sir Ellis Kadoorie and James Taggart, who had the same idea — a hotel to surpass in its standards anything yet known in the Far East. Sir Ellis died while it was still on the drawing-board. Taggart directed and supervised every conceivable detail of it. In the background of the years we are now coming to, in which Taggart figures, lay the building and furnishing of his most enduring memorial, the Peninsula Hotel.

Still, why would Chater have asked him to promote a new telephone company? The answer is, we simply do not know. Certain things are known, however. From the moment when Taggart became managing director of the Kadoorie hotel group he began to spread his wings as a business man, founding a number of companies of his own, mainly in real estate, finance, and construction. A utility company would add a civic dimension to his commercial activities. Chater knew Taggart as a hotel manager. On the rare occasions when Chater entertained outside his own home he usually did so at the Hongkong Hotel.

So far so good, but there may have been something else. Chater knew the early story of the telephone in Hongkong. He had been one of Suenson's first subscribers, and understood why the service had folded up. Though the cabal still existed in 1924, it no longer possessed unchallenged power. Other interests had come in which it had been unable to subdue or dislodge. Of these, the most significant were the Kadoorie interests, with which Taggart was associated. As the composition of the Hongkong Telephone Company's first board of directors shows, Chater's aim was that it should be part cabal, part non-cabal, with the latter predominant.

As to his inviting Taggart to take the lead, with himself in second place, this was typically Paul Chater. He delighted in being chairman of the board of stewards of the Hongkong Jockey Club — he was chairman for 34 years — because racing was an all-Hongkong affair. In any enterprise with a preponderance of Britons in it, as an Armenian he would never take the lead. Vice-chairman by all means; chairman never.

This is about as far as we shall get to understanding the position at which this chapter started, in January 1924.

In that month William Preece, senior partner in Preece, Cardew & Rider, passed through Hongkong on his way back from Shanghai, and gave Taggart the true valuation of the China and Japan Telephone Company, which was £90,000, advising him off-the-record not to offer a penny more.

Taggart had sized up the telephone company. He realized that they were determined to stay put, evidently believing that in the end the Hongkong Government would give in and let them have another licence. With Chater's agreement he went to see Fletcher in the Secretariat. He and Sir Paul Chater, he explained, proposed to make an offer to the company on a take it or leave it basis. If the company decided to leave it, they would themselves form a telephone company forthwith, and put in an entirely new installation with an automatic exchange and efficient operation. He scarcely needed to say that when this service started it would kill China and Japan Telephone, probably within weeks.

Fletcher, who had already had enough embarrassments over telephones, decided that he must warn the Colonial Office first, since the London company was almost sure to petition them against so drastic a challenge. The Colonial Office, sympathetic to the idea though with slight misgivings, decided to look at the licence.

It had disappeared. The Colonial Office did not have a copy; the Government of Hongkong had lost theirs. They could hardly ask the company to show them the original.

Fletcher was by this time again on leave in England. It was finally decided that there would be no harm in his dealing with the company personally.

The chairman of the Oriental Telephone Company was now Sir George Gibb, who had spent his life in railway administration. Between 1906 and 1910 he had been chairman of London's Metropolitan Railway, Europe's first underground service. Parker Ness was still vice-chairman of Oriental Telephone. Fletcher saw them both. Yielding nothing and revealing nothing, as usual, they proposed another visit to Hongkong. As Fletcher realized, if Parker Ness were seen there again, the uproar would be even worse than on the former occasion. Somehow he managed to convey that it would be in the company's interests if Sir George Gibb came in person; and he concluded by stating that the purpose of the visit *must* be to reach an agreement with the promoters of the Hongkong company being formed.

Gibb arrived in January 1925 with proposals which underrated the acumen, even the intelligence, of those to whom they were made. Indeed, it could even be said that his company underrated Hongkong as a whole, and had been doing so for years. It was the London view of the colonies.

The new company, Sir George proposed, would buy all the shares of the old one, taking over its assets and liabilities, Oriental Telephone to have the right to appoint two directors to the board of the new company and handle their buying. As Taggart saw it, this was the same thing again in a different form. He and Sir Paul Chater were not interested in buying the company's shares, still less in taking over its liabilities; they wished to buy it as a going concern.

Chater left it to Taggart to deal with Sir George Gibb who, in a series of long and difficult interviews, contested every inch of the way in a lofty tone of benevolent condescension, mingled here and there with an injured righteousness which wore ill when set beside his company's record.

Taggart was helped by the season. Sir George Gibb arrived just in time for the annual race meeting — five days of racing, during which Hongkong closed down for a week. This was followed by Chinese New Year, when most of Hongkong closed down for a fortnight. Sir George became so disturbed by the delay that he changed his berth to a later sailing. Agreement having been reached shortly after this, he was left for a fortnight waiting for the ship.

As with Parker Ness, however, who had sailed away in a flourish of silken threats, impervious to having infuriated — and cheated — three major cities of the Orient in as many months, Sir George Gibb, when his ship at last arrived, sailed in high tone. If he had been in a controversial mood, he wrote to Taggart, he would have presented 'in a very different light' some of his arguments in their discussions. It was benevolent condescension to the end. When the press published the letter — Sir George made sure of that — Taggart let it pass without comment.

Basic to what happened next is the fact that these were years of social unrest which was being stirred up in Hongkong by Russian communists in Canton, acting on orders from Moscow. There had already been a seamen's strike (1922) which had paralysed the port for nearly three months. At the time we have now reached — March 1925 — there were indications of a build-up to something even larger: a total strike and a mass exodus of the working population to Canton. It was to be accompanied by something larger and even more serious: a total boycott of British goods in China. All that was required was an 'incident' to give it a *raison d'être* and set it off.

The Hongkong Telephone Company was formed against this background, which goes some way to explaining certain features which would otherwise be incomprehensible. The company was to have share capital of $5,000,000, in 500,000 shares of $10 each. Half the shares were to be offered to the public; 140,000 were to be allotted to the former company as the main part of payment for the acquisition of its plant, land, buildings, and stores; Sir Paul Chater and James Taggart were to receive 25,000 shares each as promoters, the remainder going to the other directors.

A Telephone Ordinance was drafted. The law officers had before them the 1922 conditions proposed by the joint committee of the chambers of commerce, to be applied to the former company. These, owing mainly to that company's secretiveness and the lack of trust this inspired, were stringent conditions, few of which would apply to a local company with a policy of openness and the criticism of local shareholders to contend with. Having nothing else to go on, however, the law officers built on these conditions, even tightening them up.

The outcome was an ordinance restrictive in character and entirely unsuited to the financial management of a public utility. It was tabled in the Legislative Council and passed into law without anyone concerned drawing attention to its defects, either privately when it was in draft, or in the course of debate.

Under it the dividend and rate were fixed, and could not be altered without government permission; surplus funds (above the fixed dividend) had to be distributed, but only with government approval, and could not be placed on reserve; the company was not allowed to increase its capital without government permission.

A public utility cannot function properly under such conditions. This was to be a public company under a form of government control which was entirely restrictive, not promotional. The wound in the body of this was that it was governmental financial thinking, which is annual in nature, applied to public utility finance, which is forward financing. At any juncture requiring promotion, the company would have to apply to the Government. There things would break down, for the simple reason that no government can be expected to think in the same way as a public utility manager, who if he is any good is thinking six years ahead, and is consistently advising his directors to take calculated risks. That is something which no government will ever do, yet without which a utility will dangle perpetually on the brink of obsolescence. A reserve fund, denied under the ordinance, is an absolute essential. Capital needs to be increased years before it is needed — something which no government can ever understand. Fluidity needs to be on a larger scale than is normal in other commercial concerns. None of these basic public utility rules could be applied by the Hongkong Telephone Company. It was born with clipped wings.

It is not surprising that James Taggart did not draw attention to the ordinance's defects; he had no experience of a public utility. It is very surprising that Robert Shewan, one of the directors, himself the initiator of the tramways and of Kowloon's electric company, did not raise objections. Most surprising of all is the acquiescence of Sir Paul Chater, who knew more about public utilities than anyone in Hongkong, and was the architect of the financial policy of the Hongkong Electric Company, which consequently ranked as a model to all utilities anywhere (by the way, it had no licence, and was under no form of government control). Granted, Sir Paul was in the last year of his life. He was still very alert, though, as will be seen in a moment. A word from him in the Executive Council, an after-the-agenda word, and the ordinance could have been changed.

The ordinance having been passed into law, and the formalities of registration and commencement of business having been completed, the telephone company began to function under its new name, with the same staff, on 1 July 1925. It had not yet floated any shares; it had not even published a prospectus.

Two days before, the apprehended political 'incident' had occurred — actually in Shanghai — and a week later the so-called General Strike began. Apart from the handful of Europeans, nearly the entire staff left, boarded trains, and went to Canton. Company wives and outside women volunteers came to the rescue. Considering that 75 per cent of the calls were in Chinese and the volunteers were 'memsahibs' who did not even know the numerals in Chinese, it must have been quite funny. The public telephone service — suddenly rather important — never broke down.

All the same, thinking back to Suenson, and to the dismal beginnings of the China and Japan Telephone Company, if anyone had said that some Pharaoh had placed a curse on telephones in Hongkong — governmental ones excepted — one could almost have believed it. Surely nowhere in the world has the telephone had a stranger story than in Hongkong. The real and abiding curse now, of course, was the ordinance.

It soon came into play. The General Strike was initially a Russian triumph. Silence reigned along the entire waterfront. A footfall in Pedder Street could be heard in Statue Square. The Government, not understanding the exact political nature of the situation, called for Europeans to volunteer to man essential services. Response was mainly from the Portuguese and Eurasian communities; most of them were employees, and there were not nearly enough of them. After six weeks of almost total stagnation a lawyer, Dr Ts'o Seen-wan, took matters out of government hands, commandeered an office in the City Hall, and called for Chinese volunteers. To the Government's surprise, the response was overwhelming. In a matter of days Dr Ts'o had over 2,000 men, many of them from prosperous families, doing the most unlikely jobs all over the place, and a waiting list of another 1,000. The message was correctly read in Canton, where besides there was a food shortage and soaring prices. Gradually people came back.

By October, three months after it started, things were sufficiently normal again for the directors of the telephone company to issue a prospectus and float the shares. Sir Paul Chater, the acknowledged expert in matters of this kind, advised a low first call (down-payment) of $2.50 on each $10 share.

The Government, knowing nothing about stocks and shares, promptly intervened, saying that the call was far too low. (They seem to have thought that the full price should be asked.) When the board stuck to their guns, the Government insisted on a premium of $1 being imposed on every share, making it in effect a $3.50 call. Sir Paul warned that conditions were not sufficiently settled; the shares would not be taken up at that price. The board, wishing to avoid a confrontation with the Government, took note of Sir Paul's views, and submitted to the Government.

Shares were launched as the Government had directed, and of course Sir Paul was right. Of the 250,000 shares offered, only 99,000 were taken up.

Under this hopelessly bad ordinance, at the first touch of the governmental hand, things had gone wrong. They continued to do so for another thirty years with lugubrious regularity.

Eastern Extension & Great Northern Telegraph office on Connaught Road, built in 1898, replaced by Electra House in 1950. *Below*, receiving telegraph messages in London from India and the Far East, about 1910.

Shanghai's world-famous Bund in its heyday. Hongkong in the 1930s, *below*, was by comparison a quiet trading port. The Hongkong and Shanghai Bank building, seen on the right, was built in 1934 partly to inspire confidence.

Chapter 17

Fifteen Years for Five Kilowatts

THE existence of electro-magnetic waves, which could pass through obstacles such as the walls of buildings, had been demonstrated in 1887 by Heinrich Hertz. In 1894 a twenty-year-old Italian, Guglielmo Marconi, convinced that such waves could convey messages — signals at least, probably even speech — carried out experiments in his parents' estate near Bologna, conclusively proving that a telegraph message could be sent without the need for wire.

The Italian Posts and Telegraphs, to whom he presented his findings, manifested surprising and disappointing indifference. Marconi's mother was Irish. At the instance of relatives on his mother's side he came to London, where his first patent for wireless telegraphy — the first in the world — was issued in 1896. He was twenty-two.

Next year he formed a British company to exploit his inventions. Thereafter, attracting enormous public interest, one sensation after another followed. At Spezia in 1897 he transmitted wireless messages from the shore to Italian warships 12 miles out to sea. A year later he sent messages across a distance of 32 miles, from North Foreland to Boulogne. In 1899 he caused a sensation in the United States, sending wireless reports from two American vessels to New York newspapers on the progress of the America's Cup yacht race. In 1900 he patented his invention which produced the tuned wave, limiting interference (United Kingdom Patent 7777).

It was universally held that electro-magnetic waves travelled straight, and that since they could not bend to the curvature of the earth, the greatest distance over which a wireless message could be sent was not more than 100 miles, and then only from high points. On 16 December 1901, Marconi's engineers in Cornwall succeeded in transmitting to him in Newfoundland the letter 'S' — three short strokes in the Morse code — over a distance of 1,700 miles.

This was a world event. No previous scientific achievement

had ever produced so instantaneous and widespread an acclamation. The imagination of the entire civilized world was captured by it. Every newspaper worthy of the name, in nearly every country on earth, carried the statement. It read:

> London, 16th December,
> 10 a.m.
>
> Signor Marconi at Newfoundland announces that he has there received faint but conclusive signals from Cornwall by wireless telegraphy.

The Eastern Telegraph group of companies (the Pender cable empire) then made a stupendous mistake. On that very same day, 16 December 1901, within hours, perhaps within minutes, of the statement being known, they informed Marconi through his London office that one of their companies held exclusive telegraphic rights in Newfoundland, and that if he uttered another signal there he would face a lawsuit.

Dismantling his installation in Newfoundland, Marconi accepted an offer made to him by Graham Bell, inventor of the telephone, enabling him to continue his experiments in Nova Scotia, and two months later formed the Marconi Wireless Telegraph Company of America. World leadership in communications passed from Britain to the United States.

Britain thereafter did not merely fall behind. In less than twenty years Britain had dropped to bottom in the communications league. As for the Eastern Telegraph group, having made their first irreparable error, they declined over the years by way of one faulty decision after another until 1927, when Sir John Denison-Pender, the founder's son, appealed to the British Government to rescue his companies from total collapse.

This was the background to the Hongkong developments — or lack of them — which will now be related. Underlying the scene was the fact that nearly every Briton in authority, whether at Westminster or in Whitehall, mistrusted wireless, which they did not understand, preferring cables, which were part of their tradition. If Marconi could prove that wireless really did work (there was uncertainty on this point), and if the Government felt they could run it themselves, they would take it out of his hands, or else suppress it for reasons of naval defence, or to protect the cable interests. Either way, they felt the need to control wireless, whether to promote it or to impede it. In short, they did not know what to do, nor, as it turned out, did they know what they were doing.

In December 1901 Marconi's patents for wireless telegraphy were registered in Hongkong, with a view to Marconi's British company setting up a commercial ship-to-shore wireless service. The Admiralty moved promptly to have the Hongkong patent annulled on the grounds that 'an independent station could be prejudicial to the general public interest'. The use of the word 'general' in this seemingly baffling statement meant that it included an 'Imperial interest' in Hongkong 'as a Fortress and Naval Station as well as a commercial city'.

The Attorney-General, in response to prior naval threatenings (made locally) had issued the patent with the proviso that 'it be not prejudicial or inconvenient to His Majesty's subjects in general'. The use of the word 'general' being now clear, the upshot was that the patent was not annulled, but there was no wireless station.

In 1903 a Wireless Telegraphy Ordinance was passed in Hongkong under which licences could be issued for wireless operation. The Colonial Office had advised this, and it had their blessing. Eastern Extension applied for a licence to communicate by wireless with their cable repair ships. Weeks of dead silence following this application, it became clear that this was a facade ordinance. Licences might be issued in the Governor's name, but not before they had been referred to London.

The Admiralty, horrified by the dangers to naval communications which the Hongkong ordinance posed, demanded of the Colonial Office that no wireless licences be issued to any commercial company. The Colonial Office obediently directed the Governor accordingly. Sir Matthew Nathan, Hongkong's youngest-ever Governor — he was thirty-two — and one of the most progressive, felt that such an order amounted to an insult. In his reply to London he made it clear that he obeyed under duress.

The Naval Commander-in-Chief in Hongkong then approached him. The Navy was having difficulty in using their own wireless (they used Marconi's system) from a ship in the harbour — HMS *Tamar,* the headquarters vessel — because they were shut in by high land. The Admiral wished to set up a Royal Navy wireless signalling station on Cape D'Aguilar, the extreme south-eastern point of Hongkong Island, site of a disused lighthouse.

Nathan came to an agreement with him. The Navy could have the Cape D'Aguilar site, and would take over all signalling and wireless telegraphy in Hongkong, naval *and* commercial, in peace and war. Providentially there was a precedent for this: Malta, where the Navy handled commercial wireless messages. It looked as if things were moving.

The Admiralty disavowed its Hongkong Commander-in-Chief's arrangements, ordering deferment of all change until the matter had been further considered. Two years later, in 1906, they decided against a permanent wireless station in Hongkong. (Their own wireless was by then operating on Stonecutters Island.) They made one concession. After waiting nearly three years, Eastern Extension was allowed wireless communication with its cable ships, under rigid conditions: if there was the slightest interference with naval communications, the licence would be cancelled. No other licences were to be issued.

In 1910 Marconi suggested to the British Government the creation of an Imperial Wireless link. Coming from Marconi, the suggestion was found disconcerting, and was turned down. When, however, at an Imperial Conference held in London the following year, the same suggestion was made by the Prime Minister of New Zealand, it was favourably received. A committee was formed to make recommendations.

At its meetings, the Admiralty made a tremendous to-do about interference with naval communications. Extra complications arose over which of the British Imperial possessions were to have high-power wireless stations, and which were to have low-power. Hongkong was not even included in the Imperial Wireless scheme. After a high-power station had been established at Singapore, there *might* be a connection between Singapore and Hongkong. Consideration of the high-power station was then (August 1911) deferred *sine die*, so neither Hongkong nor Singapore had wireless.

During 1912 the Colonial Office gradually came round to the view that there ought to be commercial wireless stations at these two places. For Hongkong a 5-kW station was suggested. It all moved with interminable slowness. Towards the end of 1913, having bungled the tender arrangements and nearly landed themselves with a bunch of crooks, they finally awarded the contract to Marconi. But then nothing happened. It seems that between the Colonial Office and the Crown Agents someone omitted to tell the Marconi company the contract was theirs.

Throughout these years 1906–14 there ran a steady undercurrent of complaint from Hongkong's mercantile community, punctuated by occasional outbursts from the Chamber of Commerce. The Royal Navy's handling of the wireless telegraphy situation was almost indescribable. Merchant ships and foreign warships were only allowed to operate *their own wireless* for four hours a day — 6 to 8 a.m. and 5 to 7 p.m. — in Hongkong waters, and must stop at once if requested by any of His Majesty's ships, while the Government of Hongkong had to ask the Royal Navy's permission before renewing the licences of world-renowned international shipping lines of all nationalities using the port. Nothing remotely like this was to be found at any other port in the Far East. A particular cause of complaint by the shipping companies was not being allowed to send messages at night, which gave greater wireless distance.

There was another point which was worrying numerous shipping lines. A letter of April 1914 from the India Line of Liverpool to the Postmaster-General in London epitomizes it.

> Ship wireless [they wrote] has to be carried, following the recent International Conference on Safety of Life at Sea; but the expense of installing ship wireless should be compensated for by commercial advantages, such as prompt reception of wireless messages at ports the ships call at. The Masters of our steamers in Far Eastern Waters complain that the system of radio-telegraphy with which their vessels have been equipped is rendered practically useless owing to the lack of facilities for the reception of such messages on land. They refer particularly to the ports of Singapore and Hongkong, and the China coast generally. All messages for Singapore have to be transmitted through the Dutch station at Sabang, the nearest wireless station to Singapore.

The last point encapsulates the situation. The Dutch, the French, the Americans in Manila, the Japanese, all had up-to-date wireless systems. Britain was way behind.

In February 1914 another mercantile outburst was brewing. The Governor was Sir Henry May, a hard, unlikeable man, yet with long experience of Hongkong — nearly his entire career. He was not to be fooled with.

'When is the wireless telegraph station to be erected?' he asked the Colonial Office. 'It is very desirable that it be started as quickly as possible. I desire to inform the public.'

'The Crown Agents have already been instructed to place the contract with Marconi', they replied.

Sir Henry May asked for a complete set of post office forms as used in a British coastal station, and all accounting arrangements, enabling the Royal Navy to handle these.

Over the past seven years the Royal Navy had been providing naval ratings to operate the Government's signalling stations. Naval ratings could presumably be the wireless operators. Sir Henry May offered the Admiral the run of Cape D'Aguilar as a wireless station. Ten years before, it had been the Navy's first choice, giving unimpeded outside range.

The Admiral, Sir Martyn Jerram, demurred. Cape D'Aguilar was vulnerable, too exposed. The wireless station should be on HMS *Tamar* — the very position from which they had sought escape ten years before — 'since the ship is within the harbour defences in case of war'.

In that case, Sir Henry May told him, the Government of Hongkong would build the commercial station themselves, on Cape D'Aguilar.

For which he gave immediate orders. Aware that, after this abrupt exchange, if the matter was referred to London another four or five years would probably go by without anything being decided, he prudently informed the Colonial Office only when it was too late for them to stop it.

At some stage in the course of these *entretiens* it was realized in Whitehall that the Marconi company did not know they had been awarded the contract. This was rectified.

Chapter 18

Imperial Speech without Sound

THE Government Radio Telegraph Station at Cape D'Aguilar opened on 15 July 1915, with Marconi equipment and Marconi operators. At the General Post Office in town there was a Radio Telegraph Office with a choice of telegraph and telephone communication with Cape D'Aguilar. The two telephone systems, governmental and public, had exchanges at the Post Office, to which the Harbour Office signalling, lighthouse, and Observatory communications were transferred, bringing all government telegraph services for the first time under one roof.

Indicative of the topographical difficulties of Hongkong Island, personnel stationed at Cape D'Aguilar were supplied with food and drink fortnightly by sea, the only practicable way of bringing stores to them, the alternative being a sixteen-mile walk, much of it over exceptionally difficult terrain. From the beginnings of international cables, men 'in the field' had had to be prepared for a rugged life with few comforts. Wireless continued the tradition. For most of the year a major item in the supply vessel was ice. With careful use of this, the men could count on fresh meat four days a month. For the rest of the time it was hard rations. Fresh vegetables and fruit were something of a luxury even in Hongkong. At Cape D'Aguilar they were seldom seen. Mercifully they had Chinese cooks, the most ingenious of men in dealing with hardship.

The station had not been operating for more than three weeks when the Royal Navy changed its mind about Cape D'Aguilar. Certain wireless directions being screened from HMS *Tamar* — this actually meant Stonecutters Island — the Admiral required to take over the government station. Sir Henry May, who seldom smiled and was never known to laugh, was the right man to deal with this. The Great War was raging in Europe, and these being war conditions, he had to give in, but did so on terms. The Navy would run the station with the existing operators and accounting arrangements, that is, at its own cost to government profit. The Admiral agreed, it not being thought that the war would last more than another six months. The Marconi operators became naval ratings.

Unfortunately, the wartime restrictions on merchant ships in the use of wireless were so severe that the volume of commercial messages was very small. As Hongkong's Postmaster-General observed, despite the earlier public clamour for it, the station was running at a severe loss. Then, in 1916, the war far from over, the French asked if they could use the station for transmission of news of the Western Front to their compatriots in Indo-China. This improved matters, though still not enough to make a profit.

In November 1918, when the war ended, these transmissions ceased, putting the station back to where it had started. The inhibiting factor now was not wartime restrictions, but a dearth of ships. Tens of millions of tons of shipping had been sunk in the Great War, the only aspect of it which actually affected the Far East.

In 1919 Marconi for the second time proposed to the British Government a commercial Imperial Wireless link, putting forward a comprehensive scheme which of course included Hongkong. His Majesty's Government decided upon an Imperial Wireless link of its own. A committee was appointed.

This created a standstill. Marconi had by this time discovered the secrets of shortwave wireless transmission, and had invented the direction finder. In Hongkong, Sir Claud Severn was pressing for these, and for automatic telephones, meeting with no response whatever from London. Even the Admiral felt frustrated. Wishing to make wireless improvements on Stonecutters Island, to which the Navy would in due course have to return, his way was blocked by the committee, in the interests of the Imperial Wireless scheme.

Another point which Severn made to London was that what Hongkong particularly needed was a direct wireless link with Honolulu. He was privately informed that if Hongkong wanted this, they must pay for it themselves, a situation of which Sir Claud was aware. The matter was however considered in London. It would mean a high-power station, and 'until His Majesty's Government decides whether to build one at Imperial expense, would it not be better to wait?' The Colonial Office, without reference to Hongkong, decided that it would be.

So they waited; and three years later His Majesty's Government decided not to build the high-power station.

This opened the way (March 1923) to allowing the Marconi company to build such a station. The British Post Office intervened. The Marconi company would presumably need a subsidy, they remarked as a sure means of shooting down the proposal. The Marconi company did not require a subsidy, and the British Post Office knew it. They were being 'very difficult', the Colonial Office thought. In fact, by March 1924 they had produced an absolute blockage. For why, it was asked, did Hongkong need such a station? Singapore was the only important place in the region. What other station did Hongkong have any need to communicate with?

In May that year Guglielmo Marconi, in England, spoke by wireless telephone to Australia, with clear reception, creating yet another sensation. It was in part, and always had been, the sensational nature of Marconi's genius which made committee-men such as those at the British Post Office behave as they did. How they came to be overruled in the matter of Hongkong is not clear from surviving documents. Possibly they were jolted by an article in *The Times* pointing out that Britain had fallen far behind other nations in the communications field, laying much of the blame on committee-men such as they. Perhaps they simply changed their minds. At all events, Marconi's agent reached Hongkong at last in January 1925, with a view to assessing requirements.

First and foremost was the need to catch up with China, where wireless telegraphy was well advanced, in some provinces under Japanese technical control and with Japanese equipment, in other provinces American, in either case nominally subject to the Ministry of Communications in Peking.

The Government of Hongkong had regained control of Cape D'Aguilar on 1 August 1921. The Navy, from the day the war ended, had given minimal attention to it. There is evidence from the letters of sea captains that by 1920 the service was very bad indeed. From the moment Hongkong's Postmaster-General took it back, it made an amazing recuperation, launching out in an unforeseen direction, to explain which it is necessary first to explain the character of the Postmaster-General.

The senior echelon of government officials in Hongkong were members of the Cadet Service, meaning that they had begun their careers as language cadets, spending their first two years learning to read, write, and speak Chinese, usually in Canton. Working thereafter in the average government department, their Chinese usually became rusty, particularly in the Secretariat where Chinese was not used at all. There were a few, however, such as those working in the Secretariat for Chinese Affairs or as magistrates, who kept in practice.

It was an unwritten rule that the Postmaster-General must be one such. Over 90 per cent of the mail was in Chinese; most of the 'overseas' mail was with China. He maintained close contact with the Canton postal authority, often entertained visiting Chinese officials, leading a life strangely different from that of his colleagues, knowing more about events in China than any of them. In that his other principal contact was with the British Post Office, he was in a sense an 'external' man. The Secretariat knew nothing about what went on at the Post Office.

Thus the success story which now follows is not attributable to the Government so much as to the Hongkong Post Office. It began from the moment the Postmaster-General regained control of Cape D'Aguilar. A previously static situation suddenly burst into life.

By the Revolution which began on 10 October 1911, the Chinese Empire had been replaced by a Republic, with its capital in Peking, inaugurating a period of forty years which are among the worst in China's long history. On 6 April 1916, Kwangtung declared its independence of the Republic, and thereafter the province proceeded to split into two. In 1921, Canton was controlled by Dr Sun Yat-sen (ironically the founder of the Republic and its first President), while Swatow was controlled by a clique of Dr Sun's political opponents. Both places had (American) radio telegraph stations, and both wanted wireless communication with Hongkong. When the Postmaster-General, whom they knew on the postal network, took over Cape D'Aguilar, both parties knew exactly where to send to.

There was the difficulty that neither of them was registered at Berne under the Wireless Telegraphy Convention, nor was either recognized by the Central Government in Peking, meaning that no reciprocal working agreement could be made with either. Minor legalisms of this kind having been swept aside in proper Far Eastern fashion, each station collected its own charges, and the results were very effective. Cape D'Aguilar, with its wretchedly small and out-of-date wireless station, was in business. In 1922 it very nearly broke even, and in 1923 made a profit for the first time. If only it could have a more powerful station, it would soon be reaching out over South China.

The 1925 Marconi wireless installation at Cape D'Aguilar was a compromise. The Government of Hongkong was not allowed to purchase anything, not even a bag of nails, without the permission of the Colonial Office, and then only through the Crown Agents for the Colonies. In London there were the views of various committees to be considered, as too the possibility of Admiralty objections, the implications of the Imperial Wireless scheme, and more esoteric considerations. For example, Hongkong could not be permitted to have what had already been denied to Bermuda. Hongkong did get a more powerful station, but not really what it wanted and was prepared to pay for.

Within four years Hongkong was impressively in touch with the world by wireless telegram. Regardless of political turmoil, China's radio network, already substantial, continued to extend, as did Hongkong's radio contact with stations in South China.

A wireless telegram costing less than half a cablegram, in 1927 the cable companies were forced to lower their rates in China. In 1928 the Post Office pulled off its big coup: wireless link with Shanghai. Profits doubled, Shanghai coming to account for one-third of Hongkong's worldwide wireless telegram traffic. The service had quite a dramatic beginning. In its first week the cables broke down, and for three days there was no communication with Shanghai except by wireless telegram.

Radio broadcasting started in Hongkong in 1928, on the same basis as in Britain: non-commercial, to be financed by revenue from post office receiving licences. There were two wavelengths: ZBW in English, and ZEK in Cantonese, with an hour or so each week in Swatow dialect (Teochew). The English service operated from a room in the Post Office building, the Chinese from a room in Wanchai. Hours of transmission were from 6 p.m. to 11 p.m. The English service was later allowed an extra two hours on Sunday mornings for broadcasting religious services. The transmitters were installed on the top of Victoria Peak, broadcasting by shortwave to French Indo-China. Outside broadcasting hours, the transmitters were used for normal commercial traffic.

In the same year the Postmaster-General opened a wireless

school, which was to play an important part as the use of wireless extended, and later formed a separate telecommunications department under his direction.

By 1929 Hongkong had direct wireless telegram service to Bangkok, Saigon, Malabar (with connections to the Dutch East Indies, Dutch Borneo, Australasia, Europe, and America), Manila (with connections to America and Europe), and Formosa (via the Japanese Taihoku service, with connections to Japan). In China there was direct service to Shanghai (with extensive connections inland, in particular to Wuhan), Kwangchowan (a small French colony in southern Kwangtung), Canton, Swatow, and eight other major prefectural towns in Kwangtung, and perhaps most surprising of all, Yunnan-fu (Kunming), the last Chinese city before reaching the frontiers of Burma. Yunnan-fu provided connections with other places in Yunnan province, a fact which is alone enough to show how far and widely China's communications had developed, despite her incorrigible war-lords and faction-ridden politics.

To many of the more remote stations in China (those with indirect contact with Hongkong numbered well over a hundred) telegrams had to be sent on a 'receiver to pay' basis, and more or less throughout, due to the troubled conditions, reciprocal agreements were highly unusual, most of them subsisting on trust. A feature of wireless telegraphy between Hongkong and China was a conspicuous absence of cheating and default. Frequently a local political upheaval in some part of China would cause a breakdown in everything except wireless. Telegrams from Chinese radio stations such as: 'We have your money and will send it as soon as we can', would be received. It might mean waiting for months, but in the end the money would arrive.

Meanwhile, as noted above in Chapter 17, Sir John Denison-Pender had appealed to the Government to rescue his cable companies from collapse. It was the Stanley Baldwin ministry of 1924–9, characterized by profound complacency, yet even it was stirred by the appeal. Cables were Empire, part of the nation's greatness. Cables had to be rescued. Wireless, provided it was kept in second place, would rescue them.

At government suggestion, a merger of Marconi's wireless interests and the Pender cable interests was privately discussed. There was the convenience that Marconi had been eased out of the chairmanship of his own company, of which he remained a director. It meant that the key confidential meeting at which a merger was decided upon, could be between three Britons only, without the confusing presence of an Italian. (Marconi was a Senator of the Kingdom of Italy, and about to be created Marchese.)

The merger was effected in 1928. The shareholdings were combined in an investment company, Cables and Wireless Ltd. The operational company was named Imperial and International Communications Ltd. The Government appointed, on a permanent basis, an Imperial Committee to advise on the conduct of the new organization's affairs. As the first chairman of Imperial and International they chose Sir Basil Blackett, a retired Treasury official who had for the past six years been Finance Member of the Viceroy's Executive Council in Delhi. In this way the Government, without actually interfering, ensured that the interests of the Dominions would carry weight in the organization's policy decisions, and that service and help would be provided for the colonies.

In May 1929 Sir Basil Blackett, sounding the ground, suggested that Imperial and International purchase the wireless stations in Hongkong and British North Borneo, which were outside the Imperial Wireless fold. Reaction in the Hongkong Post Office was one of dread. They already had the Shanghai wavelength open; business was booming. Imperial and International, it was suspected, would prove to be the dead hand of His Majesty's Government under a new name.

The Colonial Office held an entirely different view. They liked Sir Basil's comprehensiveness. 'It would be far better', they decided, 'to have the whole thing in one huge cable and wireless concern than odd bits in small territories, always ending up with government holding the baby.'

In 1930, when Imperial and International opened negotiations with Hongkong, they found that these were not so easy as they had expected. Wireless in Hongkong was very much a going concern. Profits from it were seven times higher than the operating costs. It was not just a question of buying the Cape D'Aguilar installation, but of buying a highly profitable business. In addition, a guarantee was demanded that, within a defined date, Imperial and International would replace the existing plant with a high-power station incorporating the most up-to-date facilities to meet Hongkong's needs, which would include a radio telephone service to Japan.

In May 1930 the world-wide effects of the previous October's Wall Street crash became apparent. The Great Slump descended upon the Western world, although not as yet in the Far East, where business was thriving, and continued to be so for another three years. Imperial and International had provided the Hongkong Postmaster-General with some wireless engineers, evidently as a sweetener. When, in 1932, with negotiations still unconcluded, some of these men were retrenched by their London company and withdrawn, the Hongkong Postmaster-General advised his government that he proposed suspending the negotiations, which thereupon broke down.

The Colonial Office then changed its mind about comprehensiveness. A minute reads:

> The real facts of the situation, of course, are that there are insuperable difficulties at the very heart of the whole idea of a cable and wireless merger. New wireless services are naturally demanded by British Colonies, especially when they see Dutch, French and even Chinese wireless services throwing them into the background. On the other hand, such services can only obtain revenue at the expense of the cables, and they will accelerate the day when the Company will definitely propose to abandon their cables. When that happens, it will be extremely difficult for the various Empire Governments to resist a claim for heavy subsidies if the Company are compelled to maintain, on the grounds of their strategic value, cables for which there is no real economic justification in time of peace.

The long folly of the British attitude to wireless in relation to cables is laid bare in those words.

In April 1933 the Government of Hongkong broke off negotiations with Imperial and International which, badly knocked by the Great Slump, did not have the money to build the high-power station required. Hongkong's Postmaster-General observed that there were still four Imperial and International engineers left in the colony. If, as he assumed, these would now be retrenched, he was prepared to employ two of them.

In the same week, by sheer chance and without collusion, Singapore and Malaya also broke off negotiations with Imperial and International.

This combined verdict on the situation produced a response, albeit an anaemic one. His Majesty's Government permitted Malaya and Hongkong to go ahead on their own, and promised what help they could, except that 'no service was to be developed unless the existing facilities are demonstrably inadequate' — in other words, there would be no improvement of any kind — and as to Hongkong's request for a radio telephone link with Japan, this could not be allowed, since it would interfere with commercial lines.

The Imperial Committee, to which this decision had to be referred, were slightly less anaemic, finding that 'Hongkong and Malaya, subject to certain reservations, should be allowed to develop their own services.'

Putting it another way, 'subject to certain reservations' His Majesty's Government would desist from impeding the development of wireless in its Far Eastern colonies.

Chapter 19

Talking to America Forbidden

FORMING the Hongkong Telephone Company was Sir Paul Chater's last significant contribution to the civic and commercial life of Hongkong. He died in May 1926, when the company was still not one year old.

In forming it, he had elected to continue the existing managing agency. The Gibb Livingston chairman, Archibald Lang, was on the board of directors. Chater's aim was to achieve a balance between two commercial groups potentially in conflict. While Chater was alive, James Taggart went along with this uneasily. China and Japan Telephone, the company's largest shareholder, had invited him to be the director representing their interests on the board. While this gave balance of a kind, the situation remained uncomfortable, the more so in that Taggart was a man who, when he put his hand to anything, did not feel at ease unless he had complete control of it.

Quite early on, it was discovered that in the 'interregnum' between the two telephone companies the manager had given himself a sizeable rise in salary which he had been paying to himself for six months. Ordered to reimburse the unauthorized sums he had drawn, he refused. At Taggart's insistence he was sacked.

The manager was a technical man taking his orders from London. It was felt that a man of wider experience was required as his successor. Sir Paul Chater, striving to the end for balance, advised putting the matter to Archibald Lang, the managing agent. Taggart, with misgivings, had to concur. A few weeks later Sir Paul died.

Lang, at the next meeting, proposed as a first step the doubling of the managing agent's (his own) fees, and the appointment of the Gibb Livingston chairman (himself at that moment) as permanent chairman of the telephone company. This refinement in the managing agency system was Lang's own.

Taggart, being without the constraints imposed upon him by Chater, objected with vigour to Lang's proposals, which were

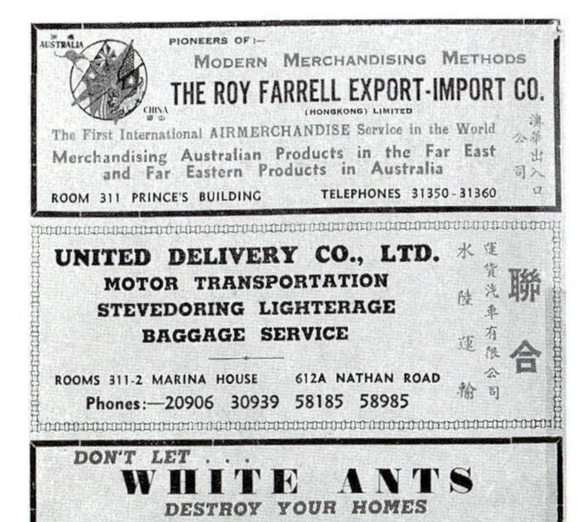

The 1946 Hongkong telephone book, 'a strange little volume', front and back. The Roy Farrell Export-Import advertisement, *left*, is the first announcement of what shortly became Cathay Pacific Airways.

Above, Electra House on the occasion of its opening on 25 November 1950. By courtesy of Cable and Wireless it became the home of Radio Hongkong, which occupied two floors. The building was later renamed Mercury House.

Left, Telephone House, Kowloon. Situated on Nathan Road, when completed in 1949 it was the colony's tallest building.

thrown out. The chair was by chance vacant, so Taggart successfully had installed in it his friend Scott Harston, a Leeds-born lawyer who was a director on several of Taggart's own companies, and a fellow-director with him on the board of Hongkong and Shanghai Hotels. Sherry, the deputy telephone manager, was appointed acting manager. James Taggart proposed himself as managing director. It was so nicely done that Lang had no alternative but to vote, like the rest, in favour.

China and Japan Telephone protested at the sacking of the manager, their former employee, demanding his reinstatement or compensation. Simultaneously trouble blew up over the $1 premium which, it may be remembered, the Government had insisted on imposing on each share. The board of Hongkong Telephone had decided to refund the premium on each fully paid-up share. The London company's 140,000 shares, which had not attracted the premium, ranked *pari passu* with the publicly offered shares. When asked to waive their rights on the premium refund, they refused, apparently expecting to receive $140,000 for nothing.

On these two points — the premium and the dismissed manager — James Taggart resolved to go to London to have it out with the company. In all the years of their existence they had been an incubus on Hongkong. Their continuing negative interference had somehow to be brought to an end, if possible in an amicable understanding.

The directors of Oriental (or China and Japan) Telephone confronted him in London in October 1926. It was an unpleasant meeting. Parker Ness, the barrister, set the tone.

'The position is', he told Taggart with a sneer, 'that Sir Paul Chater, you, and your colleagues embarked upon a venture in order to take a fat profit out of the market. Subsequent events having upset your calculation, you are now squealing.'

It was one of those situations which Chater in his lifetime had always taken care to avoid. Taggart replied that he took 'strong exception' to such remarks, which were 'a gross reflection on the members of the Hongkong board, and also on the memory of the late Sir Paul Chater, a gentleman with whom it was an honour and a privilege to be associated'.

There had been murmurings among the other directors when Parker Ness had mentioned Sir Paul Chater. They evidently felt that Parker Ness had gone too far. Not that they did not agree with him. It was just that he should not have said it.

The silence in which Taggart's words concerning Sir Paul Chater were received confirmed this.

After coming back from a meeting at which no one yielded anything, James Taggart wrote from his hotel to inform them that he could no longer be their representative. In his place they chose Lang, the Gibb Livingston man on whose obedience they could count. This satisfying quality was made clear a few days later in Hongkong, when Lang, at a board meeting, strongly advised against the introduction of an automatic exchange system.

China and Japan Telephone had never been an estimable concern. It at this stage sank to its lowest depths, doing everything it could to retard the progress of the Hongkong Telephone Company.

The new company had acquired (for 140,000 shares) a central exchange — the original one in Duddell Street — which 'operated on the Central Battery signalling principle, with Magneto calling, earth clearing, and local battery speaking'. In any other modern city in the Far East this would have been described as antediluvian.

Despite Preece, Cardew & Rider's advice to them to have an automatic exchange, and their being under an obligation to the Government to improve the plant within two years, the board — remember Lang — felt that it would be wiser to continue on the central battery principle. A magneto system would be less susceptible to humidity. There was also the difficulty that an automatic exchange would mean dismissing all the operators, and that most Chinese subscribers would be unable to read the numerals on the dials. (One is left almost speechless by this.)

The advisers had pointed out that both Shanghai and Manila had automatic exchanges. Tokyo had over half a million telephones, all automatic. Should someone be sent to Manila to make enquiries about the effects of humidity there? It turned out that no member of the board of directors (Taggart was away) had ever been to either Shanghai or Manila. They consulted their legal adviser, Sir Henry Pollock. He too had never been to Shanghai or Manila, and had no information on climatic conditions there.

The board of Hongkong Telephone had now wasted so much time that they had to ask the Government for a two-year extension of the time required for improving the system, causing considerable Secretariat annoyance.

The board had in fact begun with progressive ideas, but had

been squashed at the outset by the Government. This was the reason for the hesitancy — on which Lang could play — which we have just observed regarding an automatic exchange. It was one of the worst features of the Telephone Ordinance. The board did not dare to make bold decisions — imperative in a public utility — for fear of being shot down.

The first shooting-down — one was enough to pinion the board in hesitancy — occurred on the question of radio telephones.

Early in 1926, when the colony had settled down after the General Strike of the previous year, the board sought to open a radio telephone link with Canton, as a first step to establishing similar links with other cities in China, radio being preferred because of the unsettled conditions in the countryside.

The Government, as we observed in the previous chapter, was itself interested in radio telephones. The instant it learned of the Hongkong Telephone Company's intentions, it informed them that the word 'line' in the Telephone Ordinance referred solely to lines of wire, and not to radio lines.

Yet another defect in that lamentable ordinance had come to light. Marconi had already spoken to Australia a year before the ordinance was drafted. Surely somebody in Hongkong must have known how fast things followed whenever Marconi started something. Yet there it was. It was a law to which no one had given perceptive attention.

At the same time, and with unusual speed, the Government passed into law a revised Wireless Telegraphy Ordinance, so worded as to include telephones, and implicitly to exclude the Hongkong Telephone Company from participation. Deacons, the company's solicitors, were so appalled that Shenton, their senior partner, cabled the Secretary of State in London with a plea to prevent the King's decision 'not to exercise his power of disallowance' (the convoluted way in which a colonial law became law) until an amendment had been considered. It was too late.

The Government's behaviour over this issue came as an unpleasant shock to a new company bent on making improvements. The important thing now was 'not to stick your neck out'; and to make it worse, once a year the chairman had to face the shareholders, telling them how boldly their board was behaving.

The board of Hongkong Telephone at last sent Sherry, the manager, to England to advise on an automatic exchange. The central exchange had recently been installed in Exchange Building on Des Voeux Road Central, constructed and owned by one of Taggart's several companies, all of which had their offices in it.

In response to much complaint in the local Chinese newspapers, Hongkong Telephone managed to reduce the time it took to have an instrument installed to between seven and nine months, which was thought to be a great improvement. With 7,000 exchange lines (the population stood at 700,000) the system was nearing capacity working. There were only 45 lines available in Central District.

When Sherry returned from England, in January 1928, six months of uncertainty followed over the question whether to have the Rotary system, as in Canton and Shanghai (American Standard Telephones), or the Strowger system, also American, but which was available from Britain. Meetings were not rendered any the easier by the presence of Lang's successor, Gordon Mackie, the Gibb Livingston chairman, a man with a bullying temperament who was absolutely opposed to either system, holding that they should stick to the central battery principle. On 6 June 1928, in Mackie's absence, they finally settled for Strowger.

James Taggart then showed his exceptional personal qualities. He went out of his way to make Mackie feel at ease. The Peninsula Hotel had just opened. Taggart was very much the man of the hour. He suggested that Mackie accompany him to England on telephone company business, which Mackie did. He thus healed the breach with him, and also with Oriental Telephone in London, where changes of personnel on the board made possible a better relationship.

In another field, Taggart persistently argued the Government into amending the Telephone Ordinance slightly. From 1930 the company was permitted to hold reserve funds, although it was not allowed to put away more than $50,000 in any one year, a nonsensical restriction in regard to a public utility, particularly on a silver currency, where money had to be thrust into reserve at every conceivable opportunity.

The most important concession which Taggart won from the Government concerned radio telephones:

'If a wireless telephone system is established in the Colony, the Company shall connect its system therewith on terms laid down by the Governor-in-Council.'

This checked moves which had been secretly going on, whereby Eastern Extension, now linked with Marconi wireless in Imperial and International, would run a complete radio telephone service of their own.

The reader will recall the rift which occurred in 1896 between the cable and telephone companies. It was still there. Nothing had changed. The cable and radio side of the business being far larger and more important than the telephones, the telephone company had to watch out for the slightest movement in what was, in regard to them, a predator. Scott Harston, the chairman, and James Taggart had been watching carefully; and Taggart, with his tact and diplomacy, had — with that statement in the ordinance — rescued Hongkong Telephone from near-disaster.

The automatic exchange came into operation on 3 May 1930. This produced a new situation. For the first time the governmental telephone system — they still had no automatic exchange — was inferior to the public system. The Government not only charged a fairly high royalty on each exchange line; they also insisted on intercommunication with the public system, the service to be free of charge. Finding this a tidy economic arrangement, they took to developing their own system at the expense of Hongkong Telephone. A choice example of this occurred in 1932, when Hongkong Telephone's modern installation at the Royal Air Force station at Kai Tak — commercial flying was not yet allowed over 'the Fortress' — was replaced by obsolete government magneto telephones, enabling the Royal Air Force to have a free service.

Hongkong Telephone protested against this general tendency, though really to no purpose. The Government could do exactly what it liked with them, and the board knew it.

Typifying the company's situation, in the most important development which took place in these years — the opening of the telephone service to Canton — the initiatives throughout came from China, and not, as one might have imagined, the other way round, although care was taken on the British side not to reveal this.

American Standard Telephones were installing a Rotary automatic system for Canton. Early in 1928 their Canton manager and the Mayor of Canton, Lam Wan-kei, jointly approached Hongkong Telephone suggesting a link. The company, taking their line from Sherry, the manager, reacted with the utmost caution. Canton wanted an overhead line. The company decided they were 'not prepared to spend money on a

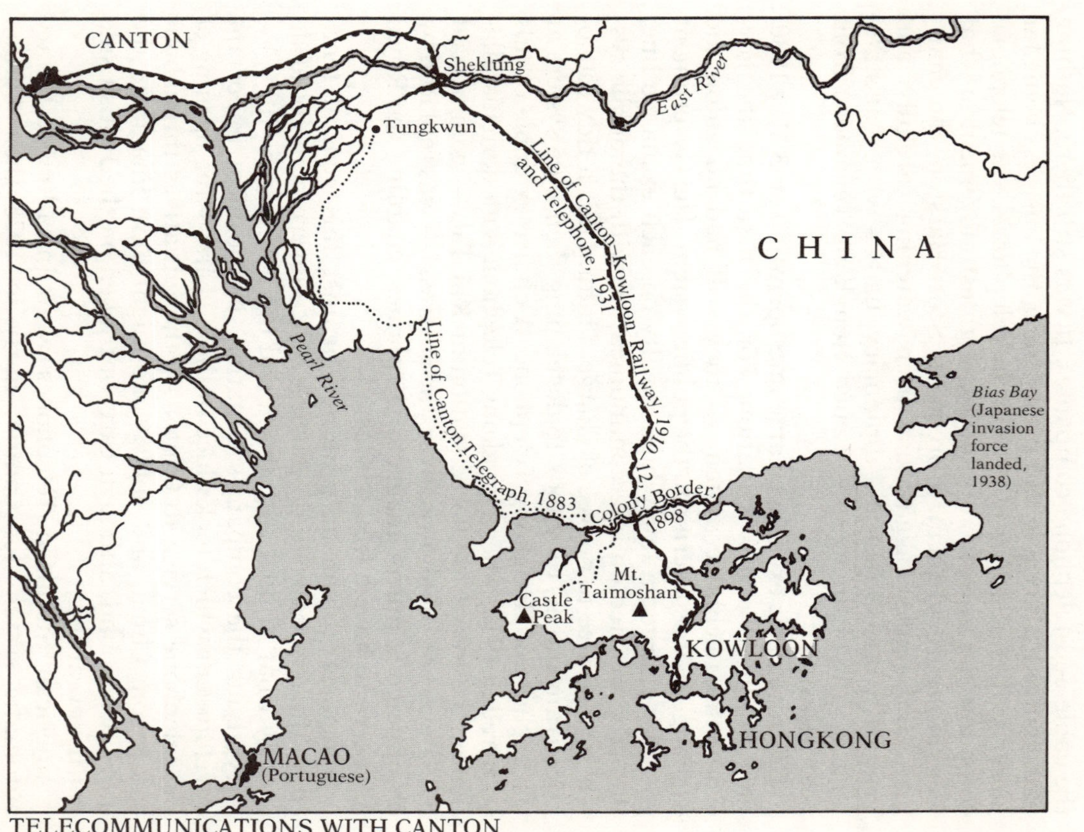

TELECOMMUNICATIONS WITH CANTON

line which, due to stress of weather and thefts of the valuable copper conductors, would rarely be in working order'. They suggested an underground line.

If they — if Sherry, when it came down to it — had known anything about the terrain, which is riverine, they would have realized that their suggestion was all but completely impracticable. The Mayor was very polite with them, not letting them know that they were demonstrating ignorance, simply persisting that the line must be overhead. After one-and-a-half years of prevarication by the company — all of it disguised to the shareholders as 'Canton having difficulties, experiencing delays' — a 30-line overhead trunk cable was decided upon, to run beside the railway.

When Sherry went to Canton to discuss the construction details, the Mayor received him delightedly, saying he hoped full agreement could be reached in a week's time, nearly scaring Sherry out of his wits. 'All aspects of the scheme, both technical and financial', he replied, would be required to be 'scrutinized very closely'. The Mayor, beginning to lose patience, pressed for a definite date by which all this 'scrutiny' would be completed. Giving six months, Sherry scurried back to Hongkong to tell Scott Harston and Taggart. A few weeks later came a formal letter from the Canton Municipal Administration. Couched in the usual very polite terms, it actually told them for heaven's sake to get a move on.

It was an unfamiliar position for Britons to be placed in. But as the British Minister in Peking had remarked only a few weeks earlier, 'Chinese can no longer be told what to do.'

The Canton line opened on 1 September 1931. Deemed first-rate, it quickly became Hongkong Telephone's best money-spinner. A great deal of Canton's business in those days was conducted at night, with the outcome that the trunk line had no slack period.

Conjoined with this, the company proposed extending service to the New Territories, by means of jointures from the trunk line to small exchanges at Taipo and Sheung Shui. The Government gave approval, yet in the accursed way in which things went wrong with the Government, they demanded that there should also be an exchange at Tsuen Wan, which was nowhere near the trunk line.

The company objected that there would be only one subscriber, the brewery at Sham Tseng, which was already connected from Kowloon. The Government then literally *ordered* them to put up an exchange at Tsuen Wan.

They did so. There were no subscribers. The thing was a complete waste of money. To justify its existence they shifted the brewery's line to it, and one other line on the verge of Kowloon. Both promptly complained at now having to pay a small charge for each call, which on the Kowloon exchange was not required.

The company should really have sent the two letters of complaint to the Secretariat, except that one could not do this with the Secretariat. One never knew where it might strike next.

It had a glorious opportunity to strike again in 1933, when the company introduced the first teleprinters (promptly taken up by the news agencies). Eastern Extension, in its usual behind-the-scenes manner, pointed out to the Government that Hongkong Telephone, with its trunk line and its teleprinters, 'would establish what was in effect a telegraph service'.

This was a good argument. By merciful accident, the Government had just broken off negotiations with the parental concern, Imperial and International, with whom they were extremely dissatisfied. For once, Hongkong Telephone was immune to danger.

Hongkong's not having radio telephone had become a disgrace. The *South China Morning Post* had drawn attention to it in 1922. It was now 1934. Hongkong Telephone already had the valuable legal safeguard that its service would be used at ground. This was not much use when there was no radio telephone set-up. They could not themselves press the Government (it could mean disaster). Instead, they asked one of the shareholders to do it at the annual meeting.

> Hongkong [he said] seems to have lagged behind neighbouring cities such as Manila and Shanghai in the matter of long-distance telephonic communication. The dignity and importance of the Colony should be a spur to our Government to push on with all possible despatch the operation of communication by telephone with the outside world.

The Postmaster-General, doing tremendous business with wireless telegrams, had let the telephone angle slip. He was not a technician; he was a Chinese-speaking negotiator and civil servant, and he was bogged down with a mass of recent international law on telecommunications, much of which made little sense in Hongkong until the commercial question of who did what became clear. In the midst of it was this enormous organization, Imperial and International, which was apparently broke.

Once again, the initiative came from China. In August 1934 the Canton Municipal Administration wrote to Hongkong Telephone, asking whether Hongkong was going over to radio telephone. They themselves were going to make a radio telephone link with Shanghai, which would give telephone communication with New York and practically anywhere. It would mean better business if Hongkong could come in on this. Why not connect the trunk line with the radio service?

For the Secretariat, when approached by Taggart on this issue, it demanded a decision of legal complexity. By law the company might operate a radio telephone service if such a radio existed. Since it did not, the prospect of the company wishing to (as it were) funnel a wire into someone else's country, and there let it take off by air, was distinctly surreptitious.

There was also the problem that Hongkong would be reaching the outside world by courtesy of the National Government of China. In 1929, however, Britain had recognized the Kuomintang as such, and since Britain, with its Imperial conferences, Imperial committees, Imperial companies, and Imperial wireless links, had spent thirty-four years achieving nothing whatsoever for Hongkong's benefit, His Majesty's Government had no one to blame but itself if it considered it had lost face. With its superb Imperial cadences, one would be inclined to say it had. In addition, it had conditionally allowed Hongkong to make its own arrangements. Permission was given.

The Great Slump had begun to bite in the Far East the previous year; in 1934 and 1935 things were getting bad — acute depression throughout. This time, when the chairman told the shareholders of delays and difficulties in Canton, he was telling the truth. In fact, for nearly three years it was a case of 'it should be ready in six months' time'.

The line opened on 1 February 1937, with the Governor, Sir Andrew Caldecott, putting through the first call to Wu Tehchen, Mayor of Shanghai. The line was excellent by the standards of the day, the service an immediate financial success. Calls to America were not so good — one had to spell out operative words — but no one complained much.

Six months later, in August 1937, Japan launched an all-out invasion for the conquest and occupation of China. The main assaults were in the North and in the Yangtse basin, aimed particularly at Nanking, which since 1927 had been the capital of China. Resistance, from Chiang Kai-shek's national armies, and from guerrillas in the North, was greater than Japan had anticipated. After a year of what was clearly going to be a long-drawn-out affair, a southern diversion was decided upon.

On 6 October 1938, a major attack on Canton began with massive air raids; 89 planes were observed on the first day over Canton alone. On the 12th an invading force was landed at Bias Bay, only a few miles from the New Territories border, and advanced inland. At 4 p.m. on the 21st, Japanese forces entered Canton and took possession.

Much of the bombing had been precision bombing, one of the earliest instances of such. The railway and the telephone trunk line had been hit at several places; the entire Canton telephone system had broken down. American and Chinese engineers worked desperately to restore the city's telephone service. Conditions there were so difficult that it was never possible to restore the trunk line, which due mainly to thefts of copper was rendered irreparable. The Japanese themselves recognized this. Over the ensuing seven years they made no attempt to reopen the line.

Thus ended Hongkong's brief experience of the radio telephone. There was to be one more, even briefer.

In London in 1934, Imperial and International Communications, evidently feeling the need to give itself a new image, changed its name to Cable and Wireless Ltd. Britain was slowly recovering from the worst of the impact of the Great Slump, while in the Far East the Great Slump had struck, the economic depression becoming steadily more grave. Cable and Wireless were on the look-out for opportunities.

The Japanese invasion of China provided them with a perfect one. The disruption of communications in China was on such a scale that in 1937–8 the Hongkong Post Office's revenue from wireless telegrams dropped from just under $1,000,000 to $126,000. Cable and Wireless, within four months of the start of this sensational fall, asked the Postmaster-General if he would like them to take over what remained of his fixed point services, which on the mainland were reduced to places in what later became known as 'Free China' — Chengtu, the capital of Szechuan province, Chungking, which became the capital of China, and of course Yunnan-fu, which became increasingly known as Kunming.

The Government, after such a revenue shock — it was not at

all a wealthy government — needed no persuading. With thankfulness on all sides, Cable and Wireless took over — obeying, it should be observed, that essential rule of public utilities, which is to take calculated risks. What they were taking over was, if not quite a dead duck, a decidedly skinny one.

The transference of responsibility took place on 1 January 1938. On the same day Eastern Extension changed its name to Cable and Wireless. The company from then on provided the Government with its entire external communications.

An experimental radio telephone link with Manila had been set up in 1935 between the Postmaster-General and the Philippine Long Distance Telephone Company. The idea was that it would probably give a clearer line to America than the proposed trunk route via Canton and Shanghai. The experiment was unsatisfactory. Special terminating equipment was needed at the Hongkong end. Business being almost stone dead at the time, the Governor did not feel that the Legislative Council would take kindly to the expense, and the project was dropped.

In 1939, after the trunk line to Canton had been knocked out, Cable and Wireless opened a radio telephone line to Manila with superb results. The Governor, Sir Geoffry Northcote, made the first call to Manila on 16 August 1939. To say that this was front-page news in Hongkong would be to put it mildly. Talking to America, the line was so clear it was a sensation. People with no urgent need to call America were doing so — Chinese mainly — just to enjoy the experience.

News from Europe indicated that war was imminent. If it broke out, it would mean in Hongkong that once more, as in the Great War, the General (or Admiral, as the case might be) could overrule the Governor.

The General did not wait for war to start. He demanded of the Governor that the telephone line to Manila be closed. Pressed for a reason, he stated that direct telephone communication with the United States presented a danger in time of war, and that present conditions warranted it. What danger the General apprehended was not clear either then or thereafter. Any normal person would have considered the line to the United States to be a safeguard.

There could be no argument, however. The Governor had to give in. On 31 August 1939, Sir Geoffry Northcote publicly ordered the closure of the service which he himself had inaugurated a fortnight before.

'It makes us feel very dead.' The newspaper had said it long ago.

In 1940 the governmental telephone system went over to automatic, the exchange being installed in the basement of the Hongkong and Shanghai Bank. Twenty years had gone by since Sir Claud Severn had first enquired about an automatic exchange.

James Taggart retired from all his managing directorships in December 1940, on grounds of ill-health. He suffered from asthma, and for climatic reasons moved to San Francisco. He had been made an OBE in 1938. With his tact, manners, and good sense, coupled with exceptional gifts of management, he had contributed much to the Hongkong of his time. Before he left, he managed to bring about some more small amendments to the Telephone Ordinance. It remained, however, a thoroughly bad law.

Stephen Grove, who joined the staff of the telephone company after World War II, accurately pointed to the kernel of the law's shortcomings: the Government's insistence on placing a ceiling on the dividend.

> Regarding distribution of profits [he wrote] the restriction on the amount which could be distributed provided no incentive for the Company to improve its efficiency and earning capacity, and caused the Board to make full use of any provisions in the Ordinance which would *reduce* the final profits.

This eventuality would have been plainly foreseen by anyone who had studied the ordinance when it was in draft in 1925. The outcome was that Hongkong Telephone grew only in exact ratio to the growth of population, which is not growth. At any time from the date of its formation, the number of exchange lines gave the same count as 1 per cent of the population. Thus, as earlier observed, when the population stood at 700,000 there were 7,000 lines. In 1940, when the population had increased to over 1,600,000, there were 16,300 lines.

On an earlier page, Hongkong was described as the uttermost end of the line. We have subsequently seen in full measure what that meant. It ceased to be even that when, on Christmas Day 1941, in the Peninsula Hotel, the Governor of Hongkong surrendered the place, and all those dwelling therein, to the military representatives of His Imperial Majesty of Japan.

PART 4

HONGKONG SPEAKING

Apart from its port, which was always
one of the largest in the world, this
place has really been sleeping
for a hundred years.
Now it has woken up.

Observation on Hongkong in the
author's notebook
1949–51

Previous page: Electronic telephone transmission was introduced in Hongkong in the mid-1960s, and with it the electronic circuit board, of which this is an example from a decade later.

Chapter 20

The Backdrop

ON that sombre Christmas Day of 1941 there was one certainty in the minds of the defeated. This was that Hongkong would one day be British again. Before that, however, for Europeans of the combatant nations lay three-and-a-half years of internment in prison camps, and for the population at large a similar period of privation as it became steadily more difficult to obtain food.

A government publication issued some years after the war explained it:

> With the Japanese occupation Hongkong died. Her free port and entrepôt trade, sole reason for her existence, had ceased, and, accordingly, everything else ceased. Possessed of no natural resources, Hongkong could only feed, clothe and employ her population through imports from overseas. These imports were no more. Unable to feed and maintain the swollen population, the Japanese conquerors encouraged a return of people to China.

By 15 August 1945, when Japan in her turn surrendered, the population of Hongkong and Kowloon had shrunk to about 320,000, roughly the same as it had been in the year 1905. The harbour was strewn with sunken ships, including ferries; there was desperately little food, no transport, no electric light, no fuel, no commerce.

The recovery which followed was extraordinary. Hongkong was on its feet faster than any other city affected by the war in any part of the world. It also became one of the most overcrowded places on earth. Throughout the year 1946 people were coming in from China at the rate of 100,000 per month. At the end of that year the population had mounted to its pre-war swollen level of more than 1,600,000.

Hongkong and Kowloon were geared at that time to house a population of about 700,000. Overcrowding in 1940, caused by an influx of refugees from the Japanese invasion of China, had been a serious problem. It was far more so in 1946. During the hostilities of December 1941 Hongkong had been subjected, from the mainland, to intensive Japanese shelling, in which 25 per cent of all residential property had been either destroyed or severely damaged. Very little of it had been restored. After the war there was an acute shortage of building materials, and these were so expensive that most property owners hesitated even to repair, let alone rebuild; and meanwhile people poured in.

Building started in earnest in 1947. By mid-1948 there was no visible sign left that Hongkong had ever been attacked in war. But even at the tremendous pace at which new buildings were going up, it would take years before there was adequate housing for those in need of it.

In the autumn of that same year 1948, it became clear to people in China that, beyond all further doubt, the National Government would fall, and that its successor would be a communist regime. Over the turn of the year 1948–9 another 600,000 people came into Hongkong from China in five months. More continued to pour in by every means available. By April 1949 there were nearly a million people living in huts made of scrap waste. There has probably never been anything like it. Building of proper accommodation was going on literally day and night seven days a week.

In that month the communists took over Shanghai. On 1 October 1949 the Central People's Government of China was proclaimed in Peking. At the end of that month, units of the Chinese Red Army reached the New Territories border where, by what was clearly a predetermined policy decision, they halted, sealing the border except for the railway entrance. Even there, trains were not allowed to run through; passengers detrained, walked across, and boarded another train on the other side.

By May 1950 the population of Hongkong had reached 2,360,000. Wanchai at that time was the most densely populated square mile on the surface of the globe.

On 25 June that year war broke out between North and South Korea. A few days later the British Government imposed restrictions on the export of goods of strategic value to China. Hongkong perforce complied. The effect was severely to curtail the colony's traditional entrepôt trade, which since the end of World War II had been flourishing.

In December, as part of America's response to active Chinese intervention in the Korean War, the United States Government

Cape D'Aguilar in the towering heat of a 1950s summer with no air-conditioning. *Below*, Hongkong in 1962, on the eve of dramatic transformation. The Mandarin Hotel is being built on the right, the Hilton upper left, and Government House, top, still has a harbour view. Not for much longer.

Charles Male, *above left*, at his home in the French Alps; Sir M. K. Lo, *above right*, on a weekend at home in Hongkong; and *below*, Richard Lee Ming-chak broadcasting from the studios at Electra House. These three men, in their different ways formidable, knocked an antiquated telephone company into shape, causing it to surpass all records.

Sir David Trench, whose war service in the Pacific was closely related to Australia's defence, speaks as Hongkong's Governor to Australia's Prime Minister, 30 March 1965. The occasion: the cable landing, *right*, at Deep Water Bay which brought Hongkong into the SEACOM radio-cable system.

New Mercury House, Wanchai, superseded Mercury House in 1972, being the largest telecommunications office building ever constructed. *Below,* the elegant cable ship *Mercury*, a familiar sight in Hongkong waters.

placed a strictly enforced and virtually complete embargo on all shipments of goods to China, throwing an economic chain around the China coast, in which Hongkong and Macao were included. In the early phase of this embargo even harmless and non-strategic goods required for use in Hongkong were refused export to the colony, and if already on the high seas were off-loaded at intermediate ports to prevent them reaching their destination.

Finally, on 18 May 1951, the General Assembly of the United Nations called on all member countries to impose an embargo on the shipment of strategic goods to China. In conformity with this, the Hongkong Government was obliged to extend its export controls over an even wider range of commodities, and to introduce import licensing on all of them. Trade as Hongkong had always known it, and which had been its lifeblood, ceased to be.

Since 1945 Hongkong industries had developed and expanded with extraordinary rapidity. The American embargo of December 1950 faced them with the most serious crisis they had yet run into in their short history. Many of them were dependent on American tin plate and other metals, American chemicals and cotton, none of these any longer available. They quickly sought and found other sources of supply, notably from Britain, continental Europe, and Japan.

At the same time their market, which was at that date essentially a Chinese market, was blocked by the insistence of the new Chinese authorities on buying wherever possible by barter arrangements. Obliged to look for wider markets, they sought them first in South-east Asia, in other Chinese and kindred communities, until, expanding the scope of their products beyond conventional Chinese items of manufacture, they broke into the markets of the Western world, with the same extraordinary rapidity which had characterized them from the start.

Hongkong, from being a pre-war sleepy entrepôt port, became a major international producer and exporter of industrial manufactures, a place of dynamic growth for which there is probably no parallel, and one of the most exciting cities in the world.

Chapter 21

Cables Devoured by Fish

No one, unless they were there, could quite imagine what Hongkong was like on 30 August 1945, when the British fleet bearing relief supplies — before all else, food — entered the harbour. The Japanese surrender had occurred with unexpected suddenness; it had taken a fortnight to provision and dispatch the fleet from Colombo. During this time emaciated Europeans had released themselves from prison camps and walked — there was no other way — for miles to their offices and installations, to find that their Chinese former employees, often in no better state than they, had done the same. There was no money to pay anybody with, but on a basis of mutual trust, work of a kind had started.

A government of starving men, members of the Cadet Service, had established itself and was giving orders fairly effectively, although on a vacuum basis, the Secretariat having lost all its records, dating back to the colony's foundation. These, with perhaps a touch of poetic justice, had been consumed as fuel. This government 'surrendered' to Admiral Sir Cecil Harcourt, who was to head a military administration for the next eight months, and — let us be frank about it — went and had something to eat.

In the course of the war the Japanese had removed mechanical equipment of all kinds — for use in Manchuria, it was obliquely said. The truth is, no one was ever able to fathom what the Japanese had been doing during the last year of the war. Much of what was discovered was inexplicable.

For Cable and Wireless it was simple. They had lost everything they possessed. Much of it had been destroyed or rendered unusable in 1941, in order to 'deny it to the enemy'; the Japanese had taken away the rest. The headquarters building on the waterfront was still there, damaged but serviceable. The Japanese had been using it, having brought in their own equipment for telegraphy and broadcasting. For three years the plant had been operated on the ground floor of Marina House, in Queen's Road Central, with antennae on the roof.

In January 1945, as protection from American air raids, the shortwave outfit was moved to the bomb-proof basement of Gloucester Building, where an emergency diesel-power engine was installed. Thence, by means of ducts to pavement level, and on stone pillars up and across the other side of Pedder Street, transmission was from the top of China Building — immediately above Hongkong's famous *dim sum* restaurant, the Café de Chine (in eclipse in those days). The receiving station was on Victoria Peak.

This set-up was handed over to the British in full working order on 1 September, almost exactly 48 hours after the fleet arrived. It was a hand-over of some moment. No one in the West knew yet whether the fleet had reached port. All the instruments capable of telling them that it had, faced east — either north-east to Osaka, or south-east to Manila. Nothing faced west.

It took three days to rectify this. On 4 September, the Japanese equipment facing west, Colombo received the first intimation of events, and relayed to London the news that Hongkong was alive again.

On the same morning as the telegraph hand-over, 1 September, the Japanese telephone manager made an orderly hand-over to Sherry, the pre-war manager, who had been a prisoner of war. Only 6,000 exchange lines were working. During the occupation, fuel supplies and telephone repairs had declined in close harmony. If a telephone was out of order, or a wire cut and stolen, that was one telephone the less. The outcome was a system in a state of disrepair which it would take months to put into order, and Sherry had had to let most of his European staff go home on recuperation leave, two of them in a hospital ship.

He had 8,000 lines working by the end of the year (in the circumstances an achievement) and printed directories in Chinese and English issued three months after the fleet arrived. In its modest way it was typical of what was taking place all around. An extraordinary vitality and determination possessed Hongkong, unlike anything known before the war. People did not say much about it, yet everyone felt it. In October, as electricity returned, an old resident caught it exactly: 'Gradually the lights came on in houses further and further up the Peak, and this was a sort of nightly graph that was watched with considerable interest.'

In marked contrast with all other cities of East Asia affected by the war, Hongkong arose from it with no electric cuts, no serious water failures, no breakdowns of the slightest importance. Everything, once restored, worked.

By the grace of God, and by a great deal of human engineering ingenuity. Confronting every utility was a serious deficiency in the supply of mechanical equipment. Sherry's experience was one of several. Having sufficient staff returned from their recuperation leave, he took his own, combining this with what his board intended to be a buying mission in Britain and the United States. In both countries he found that domestic demand for telephone equipment was so great that nothing was obtainable for export, nor was it likely to be for several years.

In fact, Hongkong Telephone were lucky. If they could get back to their pre-war 16,300 lines, there was a chance that they might avoid the drama going on in the power companies, operating on a knife-edge due to Britain's incapacity to supply machinery. Demand for telephones was not exceptional, as it was for light and power. Among Hongkong's burgeoning industries, many of them operating in makeshift conditions, were factory owners who had never had a telephone in their lives — scarcely knew what a telephone was. It meant there was a breathing space.

Cable and Wireless in London had assembled replacements for equipment destroyed in Hongkong — their station manager had managed to smuggle a detailed message to them from internment camp. These replacements did not reach Hongkong until December 1945. Prior to this, thanks to the Japanese equipment, radio telegraph lines had been opened to Chungking, Shanghai, and Manila, in addition to Colombo (for onward transmission to London), and before the end of September, with the co-operation of the Postmaster-General, broadcasting had started again.

The programmes were only audible within the bowl of Hongkong harbour. The studios were in Gloucester Building, where the Japanese had been — the upper floors were a hotel — and there they remained for five years. ZBW, the English service, calling on the range of aspiring professional talent in the British armed forces, surpassed itself in those times. A number of people who had their first experience of broadcasting on ZBW later became well-known radio names in Britain.

When in December 1945 the Cable and Wireless replacement equipment arrived, for reasons already explained it was insufficient. In addition, rehabilitation was being conducted in abnormal conditions. The place was beginning to fill up with people from China — all sorts of people, including quite a few who, after eight years of Occupied China, were out to make a fast buck, and not much worried how they did so.

Theft of wire was on a scale beyond all previous experience. Individuals and organized gangs, working by night, were digging up cables and stealing wires wherever detection was difficult. The worst-affected area was that lying between Wong Nei Chong and Cape D'Aguilar. It was often impossible to report theft quickly because telephones to police stations were not working, nor were there nearly enough police. By January 1946, when the problem was at its worst, Cable and Wireless had lost so much cable that they were in danger of running out of repair stock, while Hongkong Telephone, battling to restore their system, inexplicably chose this moment to withdraw their night fault service, making the situation worse still.

The first cable repair ship reached Hongkong waters in March 1946. The Japanese had realigned several of the long-distance cables. As the over-extended nature of their conquests became apparent, exemplified by a lack of technicians and consequent transmission interruptions, they abandoned the cables. Chinese fishermen then got into business, performing some remarkable feats. Thirty of the first fifty miles of the cable to Singapore were no longer there. Similar depredations, of varying degree, had been made on all the other main cables. It took ten months to have them all in working order again.

Underlying this rehabilitation period was a problem of a different kind. On 1 November 1945, in the House of Commons, Hugh Dalton, Chancellor of the Exchequer in the Labour ministry which had taken office at the end of July, announced the Government's intention to nationalize Cable and Wireless. Civil aviation was also to be nationalized — 'two new instalments of Socialism', Dalton described it in his diary.

For nearly a year, during which a Bill was drawn up for presentation to Parliament, uncertainty prevailed concerning Cable and Wireless' future, uncertainty which was not alleviated by the company's last chairman offering vigorous opposition to the measure, up to the last moment. The point at issue, at least in Hongkong, was whether, once nationalized, the company would maintain its character of initiative and efficiency. Of one thing there was no doubt locally: if Cable and Wireless personnel had their way, it would.

The company's position in Hongkong, however, was unsound. This had started from an early decision of Admiral Harcourt concerning the large aerial arrays required for modern radio telegraphy. There were very few suitable sites for these in Hongkong and the New Territories. The Admiral thus directed that the three armed services, the Government, and Cable and Wireless adapt themselves to the simultaneous shared use of such sites.

Among the colonial officials who deferred their recuperation leave was the pre-war Postmaster-General, Edward Wynne-Jones, who resumed his post immediately the war ended. He was a forceful personality, with political sense; moreover, he was well informed on China, and a fast operator. Inevitably he became 'chairman' of proceedings under the Admiral's arrangements. Similar sharing of the workload, with such equipment as was to hand, extended to everything in Wynne-Jones' sphere of operations. Under such impetus, this sphere quite naturally increased, until by 1946 it embraced all telecommunications.

As a means of getting things going again quickly it was a tremendous success. In the course of it, though, Cable and Wireless, deficient in plant, short of supplies, thin on the ground, and concerned about the uncertainties of the future, were reduced to a supportive role. While their personnel were making an indispensable contribution, they were doing so as individuals in a local government-run operation.

They had exclusive rights to operate fixed point wireless services. On 1 January 1947, when they became a nationalized concern, they were not exercising these rights. The radio stations on Victoria Peak and at Cape D'Aguilar, on that date as for months previously, were run entirely by the Hongkong Government, as were the special radio services for Kai Tak airport, the harbour, and broadcasting. Unless the nationalized company made a realistic effort to rectify this, the danger existed that their exclusive rights in Hongkong would die by default.

The unexpected corollary of this was that Wynne-Jones was the most determined supporter of those rights. If he had not deferred his leave, Cable and Wireless' future in Hongkong does not bear thinking about. During the military administration the Secretariat and essential departments were run by members of a Civil Affairs Unit, consisting of young men in uniform, carefully selected, and who did well, but who had no previous experience of Hongkong or China. If Wynne-Jones had not been there, one of them could easily have been running the Post Office.

One example will suffice to demonstrate the significance of this. Within weeks of war's end Hongkong was a mass of illegal wireless transmitters (the amount of Chinese human energy which streamed into the place during those months was phenomenal). The equivalent of the post office savings bank in China, desirous of conducting matters properly, sent an official to the Secretariat requesting permission to establish a transmitter in Shell House. This, to the young man who heard the request, seemed an excellent idea, affording a gesture of co-operation and goodwill. Had the Civil Affairs Unit been left to its own devices, permission would almost certainly have been given.

As it was, an approving note was sent from the Secretariat to Wynne-Jones inviting his agreement. Even at this distant date one cannot avoid a touch of sympathy for the young man over what happened then. Wynne-Jones blew up as only an old China hand could. The savings bank was part of the Chinese Ministry of Communications, with its own wireless telegraph system linking every bank branch in the whole of China. If the licence had been granted, it would in effect have given the Ministry of Communications proprietary rights on Hongkong's telegraphy with China, with a view to world rights, which was what the Ministry indeed intended to obtain, using the savings bank as a cover. As a decorative touch, the official who presented the request on behalf of the savings bank was a Scot. Knowing the Secretariat to be run by innocents, he had calculated that by going there first he might get away with it.

Having thundered this business into shape, Wynne-Jones some weeks later, when the Hongkong Government was in fact running the telecommunications set-up entirely, made a point of stating in public that the services were run 'on behalf of Cable and Wireless'. When civil government on the former colonial pattern was restored in May 1946, it shared the same intention of respecting the company's rights. In that particular year, with Dalton in mind, this required a good deal of confidence.

The nationalization, however, proved to be less radical than had been feared. The company remained intact, owning its own assets, its overseas staff employed by the company. The essential difference was that there was only one shareholder — His Majesty's Government — and the members of the Court of Directors were government nominees. In Hongkong the company's character, including its service spirit, remained entirely unchanged.

In the second part of 1947, after long and complicated negotiations mainly with Wynne-Jones' successor, J. H. B. Lee (again, a member of the Cadet Service, and the last non-professional to hold this office), agreement was reached that Cable and Wireless would take over the full range of external telecommunications, and most of the special local services, by the end of 1950. In the course of 1948 they took over the two radio stations, and by the end of that year were virtually in place on the ground, except that the administrative muddle was such that it took three-and-a-half years to disentangle.

In the midst of this traumatic reorganization, they tore down their old headquarters building of 1898, replacing it by Electra House, which with its controlled architectural style gave a new distinction to the waterfront. In it two floors were provided for Radio Hongkong (ZBW and ZEK were renamed in 1948), enabling it to vacate its 'emergency' quarters, in which it had been for five years. Following a tradition set in John Pender's day — the cable company always helping the Government when it could — the two floors were cost-free, the Government paying solely the cost of equipping them as studios — 'furnishing' was the word used. They were the best studios Radio Hongkong ever had.

In any case, it did not really matter if there was an administrative muddle in the post-war years. The point was that the public, from the first possible moment after the war, had been provided with every service it was physically possible to operate. Those immediate post-war years, particularly when compared with the experience of other cities in East Asia, were a marvel. With hindsight one might add that they were a portent of things to come.

Chapter 22

In Old Telephone Books

IN order to obtain telephone service in Hongkong during the years 1946–50 the intending subscriber had to pay a deposit of $200 and have his name entered on a waiting list, on which he might wait for anything up to two years. Telephone waiting lists were not unusual. Many countries had them at one time or another over the ensuing thirty years.

The $200 deposit was another matter. At a time when young expatriate business executives could be had for $600 a month, and when a local office worker considered himself lucky to earn $170 a month, the deposit was extortionate. Compounding the issue, to demand such deposits was illegal. The Telephone Ordinance made no provision for them.

To obtain the year's new directory the subscriber had to bring his old directory to one of the company's offices — Exchange Building, or Cameron Road in Kowloon — together with a complete set of twelve monthly receipts for the previous year, and present himself on certain set dates, between certain set hours (all of them office hours, and not including the lunch hour). In the printed notification laying down these procedures he was warned that unless he complied with them exactly he would not be eligible to receive a new directory.

This was the old colonial world's idea of a service.

Even before there was time for the post-war directors to hold their first meeting, Sherry, the telephone manager, had decided that the subscription rate needed to be doubled. In this he was undoubtedly right. The directors, when they met, supported him entirely. The scene, however, when he went to request permission for the increase, was painful. It was a meeting of two different worlds.

On the one side were the well-fed young men of the Civil Affairs Unit, bursting with health and self-assurance. On the other was this old colonial, prematurely aged by privation, as so many of them were without knowing it.

The Civil Affairs men, like their military counterparts, were not proud of what they had seen of Britain's Asiatic empire, nor of what they sensed rather than saw of pre-war colonial mores. Many of them had been shocked, for example, by Hongkong's strict racial segregation laws, about which they had known nothing until confronted with them. Sherry, all unwittingly, represented that discreditable colonial past which these young men, in the idealistic post-war atmosphere, intended to do away with; and there he sat saying he wanted double the amount of money for his telephones. It struck them as odd, in any case, to find telephones being run by a commercial company. This was not how it was at home.

Their aim, they told him (it was their mandate), was to keep the cost of living down. Utilities, under the new dispensation, should put away their ideas of making a profit. If necessary, they should run at a loss.

It sounded other-worldly, but it was the mood of the times, and the Civil Affairs men knew it. They had it from the dock companies, who were the pivot of Hongkong's restoration. The dockyards had been smashed to pieces by American bombing in January 1945, yet the companies, to the detriment of the docks themselves, had ruled out strict financing for the time being. The cry was: repair the craft, and never mind the cost. Get the port moving, and the companies would look after themselves. Sherry had completely mistimed his visit.

After gruelling argument a 50 per cent increase was guardedly suggested. When Sherry protested that this was not nearly enough, he was informed that unless he accepted this offer forthwith the military administration would take the company over. This, of course, would have been the end of the Hongkong Telephone Company. There being nothing else for it, he accepted.

This unfortunate interview was recorded as the first item in a virgin file. There were no pre-war records. When civil government was resumed in May 1946, with a Finance Branch headed by a man with no previous experience of the East, this was all they had to go on. The senior members of the Cadet Service at that time had been comparative juniors before the war, while no one in the administration remembered the days when the Government had all but begged for the Hongkong Telephone Company to be formed. The past had been erased, except for the Telephone Ordinance and its various amendments, that restrictive piece of law which peculiarly suggested that the Government had always mistrusted the telephone company. Whereas in truth it was the company's predecessor which had been regarded askance, a sentiment which by default and oversight had become attached to the new company in the ordinance of 1925.

It led with uncanny precision to the reproduction in 1946 of a situation which could be traced back to 1901: while Cable and Wireless was the apple of the governmental eye, the telephone company was mistrusted. And this had occurred with a new and different set of people, with no remembrance or knowledge of the past. Once again, it was as if some Pharaoh had placed a curse on telephones.

The company had taken impeccable pre-war measures to safeguard their finances, in Britain and the United States. The cost of post-war rehabilitation consumed their entire reserves. In 1947 it was imperative, on the face of it, to ask for an easing of the ordinance, removing the limit on the amount which could annually be placed on reserve.

The Finance Branch, demonstrating a total lack of sympathy, sent them an absolute refusal, without reasons given. For the Government's reputation, this last feature was perhaps just as well. The Financial Secretary had in fact minuted to the Governor (June 1947): 'If they go on like this, we may receive very little in the way of royalty.'

For this was what it was about, apparently. The Government wanted the cash. If the telephone service deteriorated or broke down entirely, this was neither here nor there. Their attitude was worse than anything that had happened before the war.

In another direction things went from bad to worse. Quite early in 1946 the Canton telephone authority suggested opening a line to Hongkong. Obviously the first step was to examine what remained of the pre-war trunk line, to determine whether it could be restored or should be written off. Moving about in Kwangtung was undoubtedly difficult. Nearly every bridge in the province had been broken; the roads were in a terrible state. All the same, for a fairly straightforward inspection, running beside the railway line, seven months seemed an inordinately long time. There was a sense that Sherry was delaying matters.

He had been with the old China and Japan Telephone Company. He had been with the Hongkong Telephone Company from the day it was born. He thought in old terms — he could not help doing so — and for the first time he was on his own. Before the war, James Taggart had always been there to make the decisions. In this Sherry was not alone; even the board of directors felt the loss of Taggart. He had made their decisions too.

The trunk line to Canton, and the short-lived (1937–8) international radio telephone service via Canton and Shanghai, had always been a Hongkong Telephone preserve. If the trunk line was written off, the alternative would be a radio telephone. A Hongkong Telephone preserve would thereby pass into the Cable and Wireless domain. Sherry, thinking in the past, did not care for this.

So little did he care for it that, when departing on his abortive buying mission abroad, he gave no instruction to his deputy to establish liaison with Cable and Wireless in the matter. As a result, when radio telephone was decided on, the financial negotiations were handled entirely by them. Hongkong Telephone, although closely concerned, in effect abdicated.

Cable and Wireless' first post-war radio telephone line was to Manila, the first call being made by the recently arrived Governor, Sir Alexander Grantham, on 8 September 1947. Sherry was not interested in this line; it would not be a money-spinner. Manila, which had been fought street by street in the last weeks of the war, was still in ruins. The Philippines had been declared independent on 4 July 1946 — the first colonial territory to be freed from Western domination — with foreseeable declension. This was no longer the glamorous pre-war telephone on which (for a fortnight) America sounded so near. It was an unimportant regional line, really only usable in downtown Manila.

The radio telephone line to Canton opened on 10 January 1948. The financial negotiations for it had taken eight months. China's currency was depreciating at a rate which was beginning to be alarming; the Nationalist authorities, out for every cent they could make, demanded 75 per cent of the returns. J. H. B. Lee, the Postmaster-General, helped Cable and Wireless bring this down to 66 per cent, although up to the last few days before the service opened uncertainty prevailed, with Chinese officials flying back and forth between Canton and Hongkong. With agreement at last reached, Cable and Wireless offered Hongkong Telephone 15 per cent of Hongkong's 33 per cent share.

With the service due to open in a matter of hours, Sherry was obliged to accept. He complained that he had been presented with a *fait accompli*. He had, and there would be more to follow.

Cable and Wireless had informed him months before that they intended to develop radio telephone service 'to wherever public need makes the proposition financially feasible'. Sherry having made no reaction to this, Cable and Wireless took the decisions, and Hongkong Telephone was duly informed. Not knowing what to be prepared for, they were at a disadvantage.

Instead of acknowledging his mistake and climbing down, Sherry, with intent to obtain a larger share of radio link revenue, launched into a veritable campaign of complaint against Cable and Wireless, to a point where the latter declined to have any policy dealings with Hongkong Telephone except through the intermediary of the Postmaster-General. Even by colonial standards, this was a rock-bottom situation.

The Finance Branch of the Government had relented slightly in November 1947, allowing the telephone company to raise its rate to 90 per cent above the pre-war level. On the question of a further increase, and on the more important issue of increasing their capital and reserves, the company reached a point of stultifying complexity caused entirely by the ordinance, which if followed to the letter prevented any commercial expansion by forward financing. The Finance Branch was following it to the letter. Sherry persuaded his board to go to arbitration on the issues involved.

Only when they realized that by this move they had antagonized the Government did the board wake up to the seriousness of the company's position. On bad terms simultaneously with Cable and Wireless, and with the Government, they were isolated. Really they were themselves to blame; they had been letting things drift. It was more convenient, however, and not entirely without reason, to blame Sherry. It was especially convenient in that he had reached retiring age. He was 'given leave to retire', while they sent out urgent feelers to London for a new manager.

At this point one of the three members of the Finance Branch found out about the illegal $200 deposit being demanded of each intending subscriber. This discovery had taken as long as it did because most government officials lived domestically in a kind of paradise of their own, cosseted from such terrestrial realities as telephone deposits. The official concerned had invited a business man to dinner — a rare occurrence — and the guest, being new to the place, had mentioned it.

Adding to the drama, the Crown Counsel who was consulted

by the Finance Branch leaked to the press the spicy news that the deposits were illegal. It was late in 1948. In addition to the waiting list there was now a black market in telephones. The place was seething with Shanghai Chinese who did not mind how much they spent so long as they obtained a telephone. Subscribers were being paid hundreds of dollars to relinquish theirs, which were then rewired to somewhere else. The Crown Counsel's piece of news could hardly have arrived at a more perfect moment for creating a sensation.

It was into this inglorious muddle that the new manager sailed.

The real difficulty underlying the situation was that the telephone company was a small, parochial, outdated colonial set-up which had to deal with two modern entities: Cable and Wireless — large, international, competitive, progressive — and a transformed colonial government. In Britain it was the age of the welfare state; Empire was changing into Commonwealth; colonies were being advanced towards independence. Some of this was reflected in Hongkong. Where the pre-war government existed to uphold an imperial regime, its post-war version thought in terms of the welfare of the people — with which the travails of obtaining a new telephone directory provided an illustrative contrast. The company's task was to meet the demands of an enormously enlarged population. Until it brought its thinking up to date it would not be able to meet these demands.

When the Finance Branch first challenged the company on the $200 deposits (before the leak occurred), Sir M. K. Lo joined the board of directors. Oxford-educated, the most brilliant lawyer of his time, he was a Member of the Governor's Executive Council, in which his words carried weight. His advice on the deposit issue was not to contest it on legal grounds, which was what they had been proposing to do. The deposit was far too high, he considered. It had been charged, however, as a standard requirement since the end of the Japanese occupation, and had clearly been charged in the years before it.

Sir M. K. Lo had investigated the background. A modest deposit, permitted under the ordinance, had been charged in 1937 on subscribers wishing to use the Canton trunkline telephone. The company's records showed that by the end of 1939 serious difficulties had arisen. Hongkong was packed with refugees from the Japanese invasion of China; demand for telephones had risen to bursting point. With Britain at war, there was no possibility of obtaining more plant. The burden on existing lines became so heavy that a campaign had to be launched to dissuade new applicants and to ask the public to use their telephones less. There was a formidable amount owing in unpaid bills.

Clearly, this was where the $200 deposit started. It was punitive, and it was meant to be — a crisis measure to meet a crisis situation. Equally clearly, the company would not have taken such a measure — would not have dared — without government agreement. There was nothing about it in the company's records; the Government had lost theirs. Yet the circumstances, well recorded in the company's records, left little doubt.

The Finance Branch wished to demand that the company refund all the deposits so far charged. With the prevailing black market, with hundreds of people receiving hefty sums of money for surrendering their telephones to others who had never heard of a deposit, this would have been an act of madness (all the Shanghai Chinese had other people's numbers). The Finance Branch was overruled by the Governor's Executive Council. Deposits were no longer demanded, except for international callers, and this was a modest sum. In a general sense, the entire thing was swept under the carpet.

Obtaining new directories then came within Sir M. K. Lo's purview. Was it still necessary to bring old directories and receipts at definite dates and times? It was not a query; it was a comment. A more civil and rational method of obtaining directories was introduced. No one, whether on board or management, had hitherto noticed that there was anything uncivil about the existing procedure. Inconvenience to the public meant nothing to them, nor did the minatory tone of the letters they sent to subscribers. Though they altered the wording of the letters, they saw no need for it.

Then there was the English directory, which in 1950 was a strange little volume. If one required the number of the Banque de l'Indochine, for instance, it was useless to look under B. It was under F — French Bank. Similarly, the Algemene Bank Nederland was under D — Dutch Bank. There were hundreds of such oddities. A newcomer needed a local interpreter to make sense of it. Strangest of all, the only two governmental numbers in the general list were Government House and Central Police Station — the same as in the first list of January 1890.

The English directory, which was like the script of a play in a dead language, was emblematic, even significant in its way. From 1951 onward it gradually became more intelligible, less of a puzzle.

Kai Tak airport's spectacular runway in the 1970s. It was opened in 1956 and subsequently widened. In the foreground, part of its then navigation and communication equipment.

Every telephone exchange has its own problems of design, machinery being adapted or specially made to fit odd-shaped sites, such as, *right*, this one at Causeway Bay.

The extent of telephone use in Hongkong from 1970 onward had to be seen to be believed. For those living in boats, *below*, the shore-side telephones at Aberdeen and other shelters were a help, and a call cost nothing.

Above, because there is no message rate, nobody minds someone else using their telephone; it does not cost him anything.

Public housing. The early stage in Kowloon, about 1958, *above*; and *below*, Sha Tin in 1978, all modern buildings pre-wired for telephones, the racecourse top right.

Chapter 23

Shackles Removed

THE new manager, Thomas Pugh, had been in charge of Oriental Telephone's service in Bombay over the period of Indian independence (1947) and the nationalization of public utilities in India. He established a more understanding relationship with the Hongkong Government. Rather than asking for any further increase in the subscription rate, he pursued the question of raising more capital, for which agreement was given on a fairly massive scale. With Cable and Wireless he created a relationship which led to an agreement (May 1949) on radio telephone rates. These, for the telephone company, remained at the objectionable 15 per cent rate — they had only themselves to blame — but at least it was defined and accepted with good grace. It was to remain at that rate for another thirty years.

On 1 April 1949 radio telephone service opened to Shanghai, and was extended to Nanking a fortnight later. Within a few days both cities were under communist control. These were in fact Hongkong's first telephone links with the new regime in China. Service to Taiwan opened in May, to the United States in June; then in July, at last and for the first time, Britain and Hongkong could speak to each other, via Colombo. Canton, then the capital of China, was of course the major revenue-earner for international calls.

Except for the Canton line, the reception quality was nowhere good, while the service arrangements were, by the standards of North America or Scandinavia, archaic. International calls had to be booked in advance through the operator. In order to make a call after 6 p.m. it was advisable to book it before 10 a.m. Despite an exact time being allotted, it could be taken as certain that the call would come through late. Manila was probably the worst. There an hour's delay was commonplace. Then, even after shouting at the top of one's voice, it was advisable to send a confirmatory cablegram to make sure the message had been received correctly.

From the autumn of 1952, though one had to speak loudly and slowly on an international call, it was no longer necessary to shout. On the heights of Mount Butler, in the eastern part of Hongkong Island, Cable and Wireless that year installed a powerful receiving station for long-distance telegraph, radio telephone, and broadcasting. Cape D'Aguilar became the transmitting station, with powerful new wireless transmitters installed, enabling a notable extension of telegraph and telephone service overseas. The Peak station was used for short-distance communication on very high frequency, and for air navigation aids.

Cable and Wireless subsequently gave the year 1952 as the completion date of their post-war rehabilitation in Hongkong. From the autumn of that year, Hongkong was on the electric communication map for the first time in its history. Except for Japan, then emerging from American post-war occupation, it had facilities more advanced than any in East Asia.

Beside this cool statement must be set a reminder that Hongkong lies just within the tropics. At the Cape D'Aguilar transmitting station the heat from the equipment was enormous, with only extracting fans to deal with it, air-conditioning being impossible. For four months each summer the engineers worked at a constant day-and-night temperature of 120°F. The only relief from this was to go near a door and squat down, the floor level temperature being about 105°F. This ordeal went on for seventeen years. Nevertheless, it was rehabilitation.

Would there ever be enough telephones, however? The situation in 1949 was that, with no extra plant, the company had managed to extend from its pre-war 16,300 lines to 17,634 lines — 12,984 on Hongkong Island, 4,650 in Kowloon. Britain took four years to fulfil an order — any order — for machinery or plant. Economically the world was stifled in a blanket of currency restrictions. There was almost no alternative to buying from Britain and waiting. There was in addition the fact that both board and management were inexperienced in international buying. Britain was the only country they knew anything about. There was finally the point that if they tried to buy anywhere else, and the Secretariat found out, there would be uproar and vengeful riposte from the Finance Branch, already nettled at having been overruled on the deposit issue.

Then a set-back occurred with a favourable outcome. Sherry had been advised in 1947, by a specially engaged London authority, to plan for 7,400 lines in Kowloon (17,900 lines on Hongkong Island). Sherry was a sound administrator. Realizing that with the rise in population a 7,400 projection for Kowloon was too small, he aimed much higher for development of telephone service there, ending by building an exchange at the corner of Nathan Road and Cameron Road which was the highest building in the colony.

The foundation stone was laid by Sir Alexander Grantham, the Governor, on 25 March 1948. A year later, when the building was about to go into service, the garrison was increased to 29,000 men, the largest British military force ever sent to the Far East. The aim was to defend Hongkong from possible communist attack. Privately, no one in authority believed there would be such an attack. The real aim was to instil confidence in a 2,000,000 population, most of them refugees, most of them frightened. Panic, in a city of such explosive emotions as Hongkong then was, would have been disaster.

While the troops were stationed in the New Territories, various ancillary services had to be in town. The highest building in the colony was requisitioned for a military hospital and nurses' quarters. In the chaotic accommodation conditions of 1949 there was literally nowhere else to house them. The company's plans for extension in Kowloon were postponed indefinitely, and — best of all — the Secretariat apologized. The scene was set for changes.

The misjudged telephone arbitration (it died a natural death after Pugh became manager) and the deposit affair, had made the public aware of the company's bad relations with the Government; there was a severe drop in the value of telephone shares. About to raise capital, Hongkong Telephone deferred the issue, waiting for the shares to recover. By May 1950 the market was right. On the Sunday before they were to launch the shares the Korean War broke out. The instant prediction in Hongkong was that there would be a slump. For the company it was a major crisis. If they did not raise capital they risked going out of business.

The Finance Branch, acting with astonishing rapidity, cabled the Colonial Development Corporation in London for a loan of several million to tide the company over. Perhaps, although they would never admit it, they perceived that they had been interfering with the telephone company too much and for too long.

In fact, the loan was not in the end required. It soon became apparent that there would be no slump. On the contrary, adding to the excitements of embattled Hongkong–'the Berlin of the East' — the Korean War produced quite a boom (in Singapore it produced the biggest boom in South-east Asian history). Once more the company chose its time. Half the shares were taken up by the public — not bad, considering the directors misjudged the launching date. Cable and Wireless Holdings in London took 100,000. The Hongkong and Shanghai Banking Corporation, when offered the remainder, took them at once, with the condition of a seat on the board. In October 1950 the directors were congratulating themselves on how well things had gone. Financially the company was in fair weather.

Significant of an easing of tension, the Government had invited the company to acquire by purchase the governmental telephone system, parts of which possibly dated from 1882. This transaction went through at an agreed and reasonable price at the end of 1950. By strange chance the valuation worked out at $999,999. With the imperviousness to comedy which characterizes all governments, they declined to accept the dollar which would have made it a million.

If Hongkong Telephone could be classed as a relic — which it could — it had nothing on the governmental telephone system. Valuation was rendered difficult by there being almost no plans, while such plans as existed bore little or no relation to Hongkong as it had by then become. There were almost no clues to how cables or wires reached from one place to another. They did, however, and to some unlikely places (the system was much larger than anyone thought). A goldsmith's shop in West Point, for example, had a government telephone. Evidently there had been a reason for this. The police liked to have a few telephones for the use of constables on the beat, to report fires or other emergencies. The need for the goldsmith's free telephone had long ceased, yet the telephone remained. There were several dozen such.

Most of them were manually operated, for only part of the system had been joined to the automatic exchange in the Bank basement. This particular West Point number seems to have connected directly with the Secretariat, where a prim voice said, 'Hullo' — baffling to a stray Chinese caller who in those days did not know what 'Hullo' meant.

There the Secretariat exchange still was, however, on someone's desk, with an English-speaking clerk connecting from one department to another to another — simple when you understood it.

This was in 1950, with a million refugees in the streets, in the valleys, and on the hills.

In the telephone company's state it was not a bad acquisition, all the same, in particular for the equipment and stores that went with it. It held things together, even enabling a small extension of service, while waiting for Britain's four-year delivery date.

The past started to recede in 1951, when an economic adviser was appointed in the Secretariat. The adviser had a non-government background. At last it became possible for business men to talk in the Secretariat in business terms, which with the Finance Branch it never was. Within months an important easing amendment had been made to the Telephone Ordinance. The next step — a completely new ordinance — took three years of argument and negotiation, it being in the nature of all governments that once they have grasped a controlling or restrictive right it is exceedingly difficult to persuade them to let go of it.

Under the combined forces of Sir M. K. Lo, Willie Stewart (the chairman, a shrewd old hand), and Pugh, the manager, a point was reached in 1954 that the new ordinance would leave the company 'completely free to manage all aspects of its affairs'. There remained the financial side. In December that year the Government demanded as a royalty 25 per cent of the company's taxable profits. They had expected an excessive demand, and this was 'slightly higher than envisaged'. They accepted it on the reasoning that it was 'outweighed by the advantage of greater freedom of control'.

At the time of the directors' meeting of February 1955 the new ordinance had been drawn up and, following the peculiar manner of the Hongkong legislature, had been circulated to non-official Members to ensure their agreement prior to 'debate'. In April, when the bill came before the Legislative Council, it would be passed without opposition.

The atmosphere of relief at the February meeting was pervasive. It even penetrated the secretary's minutes. At last, no longer controlled by the dividend ceiling, they could give a proper return to the shareholders. They combined this with a bonus to contract staff. Suddenly they were a real company with responsibilities to the public, the onus of efficiency resting entirely on themselves.

And so at last Hongkong had a proper telephone administration. Thinking back to João da Silva's demonstration of the instrument in 1877, it had taken a long time — 78 years, to be precise.

Chapter 24

Planners and Atmospherics

SIX years further on, Hongkong goods had burst upon the markets of the Western world, while international tourists had burst upon Hongkong. Luxurious hotels and superb shopping arcades were opening for business. The air-conditioned office had come in; people looked much smarter. Thousands of Chinese, men and women, now wore Western-style clothes.

The 1950s had been a kind of seeding period of intense activity, the type of thing described in economic reports as 'development of the infrastructure'. In the 1960s all of it came to fruition at once, and gave that impression. An entirely different Hongkong, which had unaccountably signalled its oncome in the first years of post-war population onslaught, was suddenly and emphatically there and real.

With the supreme exception of providing stability, the Government had nothing to do with it. It was done by the people. For those providing services for Hongkong — from the Government onwards — it was an unending and often worrying struggle to try and keep up with the people. Understandably, advance in the provision of services proceeded in the face of a barrage of public complaint, voluble in the press, with occasional public meetings and even demonstrations of the sitdown-and-speech variety. To some extent it was the public's substitute for politics.

Hongkong Telephone with its waiting list, which at one point in the 1960s lengthened to 57,000, caught a great deal of public complaint. Cable and Wireless, from lofty height and lonely cape, caught hardly any. Well informed locally and on a world scale, their manner of thinking ahead was impressively accurate. In less than eighteen months in 1959–60 they opened international telex service with 39 countries. It was a feat which was recognized throughout the communications world as a spectacular achievement, and it was introduced at exactly the moment when Hongkong industries became truly international and the place stood in need of telex. By 1962 they had telex to 53 countries; a Hongkong subscriber could be connected to any of 100,000 subscribers overseas. To think that this was happening in the Far East — and in the former 'Fortress' of all places — was almost incredible.

The telephone, from a far more difficult background and therefore way behind, was none the less advancing. From 1954 equipment ordered from Britain years before began to arrive. By 1957 they had five major exchanges. In 1958 they passed a momentous landmark. The number of exchange lines had always given the same count as 1 per cent of the population. In that year, with an estimated population of 2,806,000, they had 62,000 exchange lines. The number of lines was at last growing faster than the population.

The service (for those who managed to obtain a telephone) was the cheapest in the world. As early as 1951 the company had done some investigation into this. The service cost half that of Singapore, and was ten times cheaper than London. It is a curious reflection that this was an outcome of the Finance Branch's earlier insistence on keeping the rate low. The company made no mention of this, of course.

They in fact mentioned the Government very little, which when they themselves were under sustained public criticism (much of the time they were) was a tribute to their restraint. The Government was the principal obstacle standing in the way of the extension of the telephone service. This time it had nothing to do with an ordinance.

At the time of the population influx from China, despite the extraordinary atmosphere of buoyancy and come-what-may optimism prevailing in Hongkong, it was far from certain whether the refugees would stay, once peace and order were restored by the new regime in China. When more than 200,000 of them returned home in 1950 it looked like a trend. There it stopped, however, and meanwhile, in less dramatic yet steady numbers, more were coming in. In 1953 it became clear that the population as it stood was a permanent fixture. The Government embarked on the largest housing scheme ever undertaken by any government in any country. Reservoirs, schools, hospitals, all the amenities of urban life, had to be provided on a far larger scale than anything hitherto envisaged.

Much of the urban fringe of Hongkong, the wilder parts of Kowloon, and the whole of New Kowloon (north of Boundary Street, and actually in the New Territories) was a shambles of squatter huts — entire towns of them — interspersed with atrocious little factories, all of them illegal, and all in the disgusting state they were in because no one had security of land tenure. The Government's new policy had an electrifying effect here. Land prices, always high, soared beyond imagination. This, far from intimidating the public, inspired *confidence*.

Hundreds of thousands of squatters had to be decently housed, the entire shambles cleared up. This required town planning. Hongkong was developing at a fabulous rate, beyond comparison with any other place in the world. Conditions were changing so fast that it was hardly possible to plan anything. There was no static condition from which to begin a plan, as even the Financial Secretary admitted on one occasion. The result was that the town planners' plans were constantly changing. In addition, the plans were over-detailed and particularist, while in the interests of co-ordination no less than thirteen government offices had to be consulted before any decision could be made on land usage.

Hongkong Telephone, urgently in need of sites for new exchanges, was caught in this. There was no question of moving right out of town, away from planners. A telephone being a wired contrivance, an exchange had to be in or near the area it served. A high-density area might need more than one exchange. With the urban areas growing and densifying — buildings were much higher than they had been — and with telephone demand increasing with unexpected rapidity, potential exchange sites became year by year more difficult to find.

Only when a site was approved for use as an exchange, and had been pared down in size to suit the planners' plans (as of that moment) could the company's architects start work, on the basis of which machinery had to be adapted or specially designed to fit the shape of the building, every exchange having its own special problems of design.

This part of the work would in any other country have been described as fantastically quick: from drawing board to exchange operation in eighteen months. Preceding it, however, could lie anything up to three years of governmental delay. An operation which, from site purchase to exchange operation, could have been completed in twenty months was taking between four and five years.

Finally, the chairman, Harry Cleland, could stand it no longer. In 1960 he wrote a detailed and reasoned protest to the Colonial Secretary. The immediate issue was an exchange of major importance in Waterloo Road, Kowloon, from which extension was designed to the New Territories, in particular to Tsuen Wan, which in the previous seven years had grown from a group of eight rustic villages into a large and important industrial town. Between the day when the company had been *given permission to apply* for the use of the site as an exchange, to the day when permission was given for site preparation to begin, lay two-and-a-half years. (From site selection to exchange operation would be five years.) Meanwhile Tsuen Wan — not least, government departments there — was desperate for telephones.

Cleland's letter did not rest there. It went on in dreadful catalogue, the history of one exchange after another. Without uttering a word of adverse criticism — in part for that very reason — Cleland in his letter gave an eerily universal damnation of bureaucracy.

> It appears to us [he wrote] that the delays arise through the operation of the Government system rather than through the fault of any individual, since we have found your Officers to be most willing to assist us but quite unable to do anything to expedite matters outside their own particular sphere.

The Colonial Secretary apologized in person, undertaking to make amends. A year later the telephone manager was invited to see the planners' plans. These, in barest outline, shorn of perfectionist detail, showed future development schemes in and beyond Kowloon based on estimated population densities fifteen years ahead. A radical change had taken place in government thinking.

The manager of Hongkong Telephone explained what he had seen to the board: 'The expected jump of people in certain areas is staggering, as are certain of the development schemes that have been outlined.' It showed, he continued, that development in what had never been considered as 'growing' areas could 'not now be neglected for our own future planning'.

The situation between government and telephone company had in a sense been reversed. It was now the latter's turn to think boldly. But the damage which the Government had caused the company, by the unfortunate requisitioning of the Cameron Road exchange and by successive delays caused by bureaucratic inflexibility, had already been done. The company in 1960 was where it should have been in 1952. It was eight years behind the population. It was in reality more than eight years behind, because the population was itself changing in outlook, and changing rapidly. Give it another two years, and the company would not be ten years behind the population; it would be fifteen years behind. A main reason for this was that the telephone had become socially integrated into local Chinese life.

To understand this one needs to look back. When Hongkong Chinese first started taking telephone service, in 1902, no one had a telephone at home. Apart from a handful of exceptional families, the home telephone did not feature much before 1925. And then, how was it used? As late as 1949, in the average Chinese family with a telephone at home, there would be one instrument, usually in the hall, used by father, and by nobody else; and father used it for the sole purpose of being in communication with his office manager or company secretary. The telephone had no social use.

Arranging social engagements by telephone was not considered good manners, it being off-hand and casual. Even giving an informal dinner (no invitation cards) someone went to each guest's office or home to convey the invitation in person. To have used the telephone would have been considered slighting to the person invited. Nor would anyone so invited have attended the dinner.

In the course of the 1950s this and many other conventions passed away. It was a time of widespread social change, of which the key years were 1956–61. To take some random samples, in the 'new' Hongkong of 1960 workers had a weekly day off, unheard-of in 1950, when workers had two days' holiday per year. With over 40 per cent of the workforce employed in industry, and doing well, there was a little money each month to spend on inessentials. The transistor radio arrived from Japan around 1954, and following it gradually came the enjoyment of leisure. The beach at Repulse Bay on Sundays was crowded with hundreds of Chinese swimmers and holidaymakers, where formerly there had been only a few dozen Europeans. Everything became what we of the West would call more normal. The prejudice against foreign food having faded away, the sandwich and chicken drumstick lunch appeared. Small children no longer ran away screaming in terror at the sight of a foreigner. The young began to liberate themselves from parental constraints. The changes which took place affected almost everything one can think of in life, and there was no particular reason for any of them. They simply happened.

Mother now used the telephone, as did her mahjong-playing cronies. Teenagers used the telephone — and how! At the same time huge new towns, such as Tsuen Wan and Kwun Tong, had gone up — in America they would probably be called cities. Integral to the real city, when they were new they felt far out, creating a psychological need for telephones among many who in the ordinary way would have made do without. This was building up at great rate to a formidable demand for telephone service, with the company mired in a morass of governmental indecision, and unaware until 1961 — when the planners showed their plans — what they were really up against.

There had been, over the years, some alarming moments in the utilities sphere. Reservoirs were in general the gravest, and these were a government concern. On the commercial side, the telephone crisis was in a class of its own, alarming in the sense that numerically — whether in relation to people, exchanges, instruments, or money — it did not appear to have a solution, while all the time, with every passing month, demand for service inexorably mounted.

One thing was clear to anyone with a head for figures. Expanding at the fastest possible speed would not be enough. The insoluble could be met only by achieving the impossible. A colossal make-or-break effort was required to draw level with people and time. Risk would need to be taken to a degree which few utilities anywhere have had to face.

A black market in telephones then surfaced. Unlike the earlier black market of 1948–50, which had vague pretensions to being *sub rosa*, this one was almost overt. Finding a telephone broker was no problem. People aimed to be connected for $3,000, which for much of the time was a kind of standard price. Difficult jobs rated anything up to $10,000. At a Legislative Council meeting in 1962, Sir Y. K. Kan, the senior Chinese Member, voiced strong criticism of the company, converted next morning into front-page headlines.

At the same meeting there was much condemnation of governmental slowness and red tape. A mood of irritation was abroad. Criticism of the telephone company, mainly in letters and articles in the Chinese press, with occasional blow-outs in the English press, went on persistently for two years, as did the black market.

There was a strain of irrationality about it. Water rationing, an annual feature of Hongkong life since the end of the war, was severely imposed due to failure of the rains in two successive years. In 1964, when public irritation with the telephone company reached its highest pitch, water was restricted to four hours every fourth day, which in the tropics is not amusing. People were venting their spleen at anything that offered. Part of the public's annoyance over telephones was a redirected criticism of heaven for not sending rain.

In that year the company had 140,000 exchange lines for a population of 3,526,500, and a waiting list of 38,000. Were it not for the interminable governmental delays in any matter relating to land, the number of lines could have been more than double that figure by then. Yet despite these difficulties, the number of lines had risen from 19,000 to 140,000 in ten years, which in most countries would be seen as an estimable growth record. The Hongkong public would have none of it.

If the directors, as a means of defending themselves, had given the public some examples of the Government's hopeless vacillations over land usage, there can be no doubt that the public wrath would have been deflected 'up the hill'. Yet at the end of that year, when the directors engaged a public relations consultant to bridge the gap of confidence between company and public, the consultant was instructed not to touch upon this side of the matter. The directors deliberately set aside their most effective weapon of defence, considering it 'prudent' to do so.

They had in the meantime engaged a London firm of consultants to advise on management, staff, and training, and invited the British Post Office to give technical and planning advice. The British Post Office sent a team who gave them a report which demonstrated an astonishingly accurate judgement of Hongkong's unique and unprecedented conditions. The size of the urban area had doubled in five years, after all, and was all the time enlarging with breathtaking rapidity. The team estimated that there would be a requirement for 1,100,000 exchange lines in the year 1980, sixteen years ahead, advising them to proceed with this aim in view. It was a forecast which proved to be remarkably close to actuality.

Pugh, the former managing director, had died in 1957 after a few days' illness. In these unexpected circumstances, the next in line on the technical side had taken his place. It was not a wise decision. It led to lax management, a black market, and a good deal of talk of management corruption, lowering the company's repute, which had never been high. To replace the incumbent in 1965, the directors decided on 'a top-flight administrator with an engineering background to be recruited from the industrial field at home and, immediately under him, a Chief Engineer'. The latter, they explained to the British Post Office, 'should be a senior telephone engineer, with the ability and experience to direct and supervise the implementation of the development scheme you have outlined for us'.

The directors, in short, had lined themselves up for change, above all in having an administrator, rather than a technician, at the head of management. Meanwhile, the Hongkong public knowing nothing about any of this, the telephone hubbub continued.

The public relations consultant, who started his work for the telephone company in January 1965, improved the situation with alacrity, and it was not a day too soon, because in March that year a spectacular telecommunications event occurred, causing the telephone waiting list to shoot up to 57,000.

The cable, long relegated to the status of an intriguing relic of the nineteenth century (all messages of importance went by radio), had made a come-back. It had been discovered that long-distance telephone messages could be sent by submarine cable. By means of repeaters placed every so many miles along a submarine cable, the sound of the voice could be heard over an indefinite distance, devoid of atmospheric disturbance. The invention was perfected almost simultaneously in 1952 by two scientific teams, one at the Bell Laboratories of America, the other under Sir Reginald Halsey, Director of Research in the British Post Office. With the inauguration of a round-the-world radio-cable telephone system, Britain was restored to the front rank in the telecommunications field.

The first phase in this system, between Britain and Canada under the Atlantic, opened in 1961. The second phase, between Vancouver and Sydney under the Pacific, opened in 1963. Thereupon, early in 1964, in Hongkong, all important telegraph, telephone, and telex circuits with Europe were transferred from radio to the SEACOM radio-cable system, via Sydney, bringing about a decided improvement in efficiency. In November that year Cable and Wireless, who had been in the scheme with the British Post Office from the start, laid at Deep Water Bay the shore end of cable for the next phase, which was to link South-east Asia directly into the system. This was to be the dawn of a new age.

The reader will learn with relief that it was a dawn in which the Cape D'Aguilar engineers, with their hellish 120°F summers, shared. In 1962 the transmission had been totally enclosed, with air blowers sucking the air out, which was an improvement. With the laying of the Deep Water Bay cable, and the coming of the new age, Cape D'Aguilar was at last air-conditioned.

On 31 March 1965 the third phase of the radio-cable system opened. It linked Hongkong, Singapore, and Jesselton with the Pacific cable telephone. Because of the stunning improvement in clarity of speech, and in part due to the semi-automatic system introduced, the number of calls from Hongkong to Singapore rose by 73 per cent in the first four weeks. Albeit these were regional calls between relatively small populations, it was the most outstanding public response in the entire round-the-world operation. In Hongkong, public use of *all* telecommunications services increased correspondingly, making it even more unusual. As for demand for telephones, the 57,000 on the waiting list ceased to mean much. The real demand was beyond enumeration.

The new general manager of the telephone company, arriving three months after this, could find the situation in a state when it was — how should one say? — crisp.

Chapter 25

Millions and Much Clearer

THE broker and dealer in international commodities had for years dreamed of having a 'book' which could go daily round the globe, keeping pace with the sun. With the round-the-world radio-cable telephone and telex, he attained it. To operate in this manner he needed for his business a place which was awake when he himself was asleep. Whether he was in North America or in Europe, the first place which came to mind for this purpose was Hongkong. Although he had probably never been there, he knew about it through the stir caused by the advent of Hongkong manufactures in Western markets.

There it was, Western-geared and speaking English, with people of manifest efficiency and boundless energy, with a staid but solid colonial government which since 1945 had brought it with unruffled assurance through twenty of the most tempestuous years in the history of the East, where there had been wars, insurrections, rebellions and revolutions almost anywhere one looked. For the global brokerage and dealership, for the 'book' which never went to sleep, Hongkong was the place.

Within months they were arriving with intent to install themselves, forming partnerships or establishing their own firms. A type of business of which Hongkong had no previous experience, and which not even the most visionary prognosticator could have foreseen in the colony's commercial life, swept the place forward into a new and enriching phase of prosperity, on such a scale as to make even industry seem small.

At the same time, and due to the expanded opportunity which the round-the-world radio-cable afforded, the global dealership itself changed. Starting probably in Chicago, and quickly taken up elsewhere, the multi-discipline global dealership came into existence. This dealt not solely in commodities, but in anything which behaved like a commodity in the market sense, most particularly in currencies, as well as in financial futures and oil

Satisfying two related addictions, Hongkong Telephone's optical fibre cables link the Royal Hongkong Jockey Club's tote and big screen at Happy Valley, *above*, and provide vital communications for the stock exchange, *right*.

A satellite dish under construction at the earth station provided and operated by Cable and Wireless at Stanley, on the ocean side of Hongkong Island.

Left, the ubiquitous telephone: five men in discussion.

futures. International finance institutions installed themselves in the colony.

This of course meant that banks had to become sleepless, keeping pace with the multi-discipline global 'book', and the sun. In ten years Hongkong was well on the way to being one of the four largest banking centres in the world, a development which no one could have envisaged in their wildest dreams during, let us say, the embargo years of the Korean War.

An arresting feature in all of this was its cause. Cables — indeed, telecommunications in general — had always been the servant and aid of commerce, and no more than that. They never *caused* anything in commerce which could not just as well be done by letter. In this case the round-the-world radio-cable, and nothing else, actually *caused* a momentous alteration and enlargement of Hongkong's corporate life, very shortly followed by a similar development in Singapore. It was as if the servant turned and said, 'Master, follow me. I show the way.' And the master followed.

Equally arresting was the sudden dynamic reaction in Hongkong, already noted in the previous chapter, when the radio-cable telephone began to function. The reaction was similar in Singapore, though less startling than in Hongkong.

The reason for this response, more emphatic there than in any other part of the world system, was that it had activated a nerve connecting with what, twenty years later, was starting to be called the Pacific Century — the twenty-first century, of which the Pacific and its rim countries would be the arena.

Historians in the future will almost certainly date the beginning of the Pacific Century to the year 1965, when Hongkong and Singapore were connected with the Pacific radio-cable, and the Far East, in consequence of what immediately followed, began to think of itself in a different way, a twenty-first century way. Not readily discernible for another fifteen years, this was where it began.

Amid this extraordinary extension of Hongkong's corporate activity, the telephone company, in the most impressive performance ever recorded of a Hongkong commercial undertaking, caught up with the public to a point where by 1970, after five years under the new general manager, a telephone could be had on demand in most parts of the urban area, while the system itself had become mechanically one of the most advanced and sophisticated in the world.

Charles Male, the general manager, was forty-nine when he took over, with a background of unbroken success in large-scale international trading. A Wing Commander in Fighter Command during World War II, he first made his mark, when he was in his twenties, with John Holt of Liverpool, managing their trade in French West Africa, returning to them after the war to manage their shipping. From 1950 he built up the African Lakes Corporation of Nyasaland into a major concern — too ripe a plum to avoid being 'acquired' on that country's independence. Since 1958 he had been managing director of three related international trading associations in Kenya, Tanganyika, and Uganda, which between them formed the largest commercial corporation in East Africa, with an annual turnover running into tens of millions of pounds. This too had an uncomfortable encounter with independence, creating the situation which immediately ensued, when everyone in London who was advising Hongkong Telephone — including the British Post Office — told them to invite Charles Male to be their manager, since an opportunity to obtain the services of such a man would not come again.

Charles Male was distinguished by directness of thinking, a natural sense of command, and a cosmopolitan manner, of a kind which Hongkong seldom met. After he left, nine years later, he was described as having been 'dictatorial, unapproachable, hard, and strict', observations similar to those of a regiment when ordered to tidy itself up, and which as strictures, in the extremity in which Hongkong Telephone found itself in 1965, ranked with condemning a surgeon for wielding a knife. The directors had asked for a top-flight administrator, and they had got one.

With his chief engineer, R. G. Gaut, he perceived that in the matter of technical advance it would be economically unsound to await the results of new inventions in other countries; they must be ahead of other countries. It would lighten the financial burden in the future. In 1966 digital transmission was introduced, whereby actual speech was converted and transmitted by electronic pulse. It allowed a more efficient use of cables, and produced another remarkable improvement in sound. The plant was supplied by the Iwatani company of Japan, where the system was just coming into use. In West Germany, Italy, and Belgium it was in an experimental stage. With it, Hongkong Telephone stepped into the forefront of the business, years ahead of the rest of the world.

In Europe that year Male enquired about an electronic exchange, to find that there was no possibility of such equipment being on the market for another four or five years. Instead, he turned to a semi-electronic exchange system being developed by Siemens of West Germany. This was a calculated risk, he informed the directors. The British Post Office, when consulted, strongly advised using their own system, which was not yet ready; it would be cheaper. But it meant waiting. Charles Male cabled the West German Minister concerned, asking for an opinion on the Siemens system. He received a personal reply by return, giving assurances. This Olympian manner of proceeding helped to settle matters. The directors took the risk.

The first semi-electronic exchange came into service in 1970. Described as experimental, it proved itself at once. Further orders were placed, and the system was extended, once more putting Hongkong years ahead of other countries.

Meanwhile Hongkong Telephone was experiencing the highest growth rate of any telephone system in the world. Charles Male recognized at once that he had staff whose capability was unique. He set them targets on the number of exchange lines they could connect in a given time. The first target was almost impossible; they surpassed it. The second was definitely impossible; they surpassed it — just. The cost of living was rising; salaries were going up. The staff were not victimized. It was a challenge they were glad to accept.

Under good management, efficiency of this kind becomes self-generating. The staff in 1965 were already unique in the telephone business. Each year they became more efficient. In 1970, in the sixth year of the Male management, it took 8.6 Hongkong Telephone personnel to run 1,000 telephones. This compared with 12.5 in Singapore, 26.4 in New Zealand, 49 in Malaysia, 95 in Ceylon, and 110 in India. Seven years after this, Hongkong Telephone's 8.6 had enhanced to 6.6, and was to enhance still more.

The targets were started shortly after Charles Male's arrival, and their effect was immediate. A record 36,000 lines were connected in 1965. For 1967 an 'impossible' target of 45,000 was set; 45,113 lines were connected. This was the year when local communist elements attempted to stage a version of China's Cultural Revolution in the colony, with ugly demonstrations and bomb incidents. Thanks to cool-headed leadership in Government House, the superb self-confidence and restraint of the Hongkong Police, and the silent disapproval of

95 per cent of the population, the attempt failed. It requiring only a 1 per cent communist element to overthrow a government, no one could be indifferent. At the height of the disturbances, which on account of time-bombs being inconspicuously left on main streets put every person's life in danger, the telephone staff were installing new lines at the rate of 5,000 per month.

Applications for telephones were coming in at the rate of 6,000 per month, however. Public response to improved sound quality was remarkable. It could be seen at its clearest in the New Territories, where the market towns and at least one of the islands — Cheung Chau — had been connected between 1955 and 1958. Reception between one town and another was not good, and response was low. When the service suddenly improved, response came by report, among people who had perhaps used a telephone — someone else's — only once or twice in their lives. In this manner the social response came evenly across the board. This had never happened before.

Hongkong Telephone were advertising; many who responded did not buy newspapers. Wireless television started in 1967 (wired television, by the Rediffusion company, had come in ten years before); not many had a set, while local television advertisers were fumbling and not yet effective. The real medium of response was word-of-mouth report, and its message (an outcome of improved relations between company and public) was simple: Hongkong Telephone was helpful — which it never had been, except for the rich and people with influence. This change, of social significance affecting the daily lives of hundreds of thousands of people, passed almost unnoticed in town; in the country it was plain and real. A last vestige of old colonial attitudes vanished. Telephones were for ordinary people.

By December 1969 Hongkong had 500,000 exchange lines. Only 33 countries in the world had more than this, Japan being the only one in Asia. In the same year a satellite earth station was opened by Cable and Wireless at Stanley, on the ocean side of Hongkong Island, operating to a Pacific Ocean satellite. Two years later a second earth station was opened, operating to the Indian Ocean satellite. These stations, introducing an alternative medium of transmission, more than doubled Hongkong's communication capacity. Cable and Wireless had already (1967) introduced a tropospheric scatter system between Hongkong and Taiwan, giving enhanced communication quality, and the same was done in due course with the Philippines, by then internationally linked once more.

Each of these improvements was detectable on the telephone. Locally — again, it could be seen better in country than in town — the most sensational improvement occurred when Hongkong Telephone connected the New Territories, including the islands, along microwaves, using digital transmission. A technical advance completed in 1974, on the islands it was astonishing.

The islands had always been peculiarly cut off, each with a complete social life of its own, related only marginally to the Crown Colony. On Cheung Chau prior to 1958, when there were no telephones, to send a message to Hongkong Island (the Crown Colony) and receive a reply, took the messenger the best part of one working day. From Tai O, at the western end of Lantao Island, a similar operation took three working days (6 a.m. on Monday to 6.30 p.m. on Wednesday, for example). When radio telephone was introduced on Cheung Chau in 1958, the main street on the island cracked with the sound of voices yelling at their business associates on Hongkong Island. An international call was a hopeless proposition.

Cheung Chau was the first place in the New Territories to have the microwave telephone, in 1967. Noticeable at once was the relative quiet on the main street, where people spoke in jolly voices of normal pitch to their cousins and confederates in California, while almost whispering to their associates on Hongkong Island. The same facility came to the other islands with a similar outcome of social significance. Being able to talk at any time with ease to friends on Hongkong Island, people started coming into town more. Big-shot islanders could be seen giving luncheon parties in the 'best places'. No longer were they outlying island bumpkins.

These developments in the telephone system had been financed on borrowed money — bank overdraft and foreign loans. While it is questionable whether this was fitting for a commercial public utility, it was unquestionably the worldwide style of business of the 1960s. The directors, under the able chairmanship of Richard Lee Ming-chak, supported in particular by the formidable P. Y. Tang, *doyen* of the Shanghai industrialists who led Hongkong into the markets of the world, aimed for a fifty-fifty division between shareholders' money and loans, with a view to the company becoming ultimately self-financing.

The overdraft was on the Hongkong and Shanghai Bank, the loans from several European countries, including Britain, although after 1967, when Britain devalued the pound, continental loans afforded sounder prospects. Charles Male, with his long experience of international commodity finance, handled this brilliantly. On more than one occasion he negotiated loans with five-year financing periods two or three days before interest rates changed adversely, bringing acclamation from the board of directors.

From 1969 onward they were paying back their debts, yet incurring others, and were not really moving towards self-financing at all. With the dynamic pace of technical advance being pursued by the management, and which was held as bold and indispensable, there was no immediate way out. The pace of advance (and of borrowing) was quickening, moreover, borne forward by its own momentum. The fifty-fifty basis of shareholders' and borrowed money imperceptibly became a general aim; it was not an actuality. The Financial Secretary, Sir John Cowperthwaite, was kept fully informed by Dick Lee, the chairman, and showed confident unconcern. For a public utility the situation was extraordinary, yet so too were the circumstances. Given one or two more years, and they would be over the hump of the debt bridge.

Charles Male warned the directors in July 1973 that the overdraft would have to go up to $225m. by the end of the year, and to $300m. by the end of 1974. When he left the colony, early in 1974, he advised that the next step must be to raise the shareholders' contribution.

Hongkong was in the midst of a colossal stock-market boom. Central District was pandemonium, and had been for months, telephone usage being three times higher than the previous highest peak. Singapore and Malaysia were likewise gripped; every stock exchange in the East felt it. The overseas service of the BBC found that at certain times of day their listenership dropped by several million owing to local radio stock-market reports, to which the greater part of entire populations were listening. Hongkong's post-war stock exchanges having consistently been first cousin to the Macao casino, the colony outdid all others in excitement and improbability.

At a special meeting of telephone shareholders in April 1974 Dick Lee announced that the company's capital was to be doubled, to $700m. Paying tribute to Charles Male, he explained that since 1965, under his management, the company's value had increased by more than 350 per cent, while its growth rate of 445.7 per cent was, among countries with 500,000 or more telephones, the highest in the world in the period 1963–73. Japan came second, with a 264.4 per cent growth rate.

It is worth noting parenthetically that the government planners during the same period had reduced their planning time to about two years per telephone exchange. The exchanges opened under the Male management gave a combined count of over fifty years of governmental delay.

Then, in the space of twenty-four hours, the stock-market boom evaporated, followed by sympathetic evaporations in Singapore and elsewhere. The number of applicants for new telephones in Hongkong dropped dramatically. Costs in general, including salaries, had risen markedly; the company's profits had fallen slightly; there was a temporary cash-flow problem. The Bank, which traditionally regarded itself as godfather to the company, felt it should have been consulted in advance on the company's most recent Swiss franc loan, and declined to increase the overdraft. The Government, which was about to launch the construction of new towns — in reality, cities — on the mainland side of the Kowloon hills, drew in its horns, ordering a suspension of activity. The telephone company, which had placed equipment orders to meet the needs of the new towns, and raised loans enabling them to place these orders, could probably cancel the equipment but not the loans, which would have to be repaid, and the money was simply not going to be there. The petroleum-exporting countries, acting in concert on a withholding basis, quadrupled the price of oil. Drastic economies on the use of oil were imposed. Every third street lamp remained unlit; neon advertising signs were switched off or limited to hours; the price of everything using oil would have to go up.

The bus and ferry companies, the power companies, with oil on order at the old rate, would be able to keep to their existing fares and tariffs for a few weeks or months. For the telephone company — though it had nothing to do with oil — the price had to go up at once. Application was made to the Government to raise the subscription rate, and the public learned of it.

The clamour which broke forth was more strident than any which a Hongkong utility company had ever had to face. Once more, as in 1964 with the extreme water rationing, they had been obliged to ask for a rate increase in a bad year, when the public temper was frayed. The unparalleled improvement and extension of telephone service was forgotten. The sense of the company being helpful evaporated in twenty-four hours, like the stock-market boom. It was replaced by a rain of acrimony pouring forth from press and radio. Demonstration gatherings were held in public parks.

As Hongkong's 'politicians' (as they were called light-heartedly) put the case, raising the telephone rate was the first move by big business to foist artificially high prices on the public in hard times, using the oil crisis as an excuse. While this may have served as a warning to such elements of big business as may have been toying with such ideas, to aim it at the telephone company was off-target, nor did the 'politicians' know about the complicated financial position which had arisen to place in jeopardy the company's tremendous achievement when it was near the point of triumph. Even with the increased rate, which was small, they still had the cheapest telephone service in the world — by far the cheapest. In Singapore, the second cheapest, the rate was *double*. As for the complaints about the service which were daily streaming forth, there was the same quality of irrationality as in 1964. The company was indirectly being blamed for the stock-market failure, by holders of worthless scrip.

On that former occasion the Government had appointed a watch-dog committee to monitor the telephone service in the interests of the public. It was one of those cosy, fussy committees reflecting the parish pump aspect of Hongkong which the colony never quite threw off. Charles Male did not exactly ignore them; he acted without them. As an example, when he informed them that he proposed having area directories — Kowloon and Mainland, Hongkong and Islands — on the same basis as New York (Manhattan, Queens, and so on), they said that this would cause grave inconvenience to the public and must not be pursued. Next thing they knew there were area directories. Male had given the orders at the same time as informing them.

Being disregarded in this manner, as well as being proved wrong (the area directories were a great improvement), on this and numerous other matters, had not endeared the general manager to such a committee. He had left the colony on completion of contract, but they would now get their own back. Taking the public complaints seriously (most of them were trivial) and drawing attention to the company's financial policy and indebtedness, they demanded a Commission of Inquiry into the company's affairs.

The announcement of the Commission — one of the judges was to be chairman — at least caused public clamour to simmer down. The company, at formidable expense, engaged a Queen's Counsel from London to present their case. As the new general manager, F. L. Walker, observed: 'If the public is critical of the Hongkong Telephone Company, they should remember the French proverb: "This animal is very wicked — when it is attacked, it defends itself."'

The Commission sat in February 1975, the inquiry lasting nearly five months. By the time it reported, Hongkong had made a spectacular recovery from the oil crisis and the stock-market crash, business was booming, the number of applications for new telephones had shot up again, the Government had resumed its development of new towns beyond the Kowloon hills, and the company no longer had a cash-flow problem. The general manager, with international financial experience similar to his predecessor, had brought the complex issues involved to a point where, meeting all their obligations, there would be an after-tax profit the following year, a sign to the shareholders — there were thousands of them — that there was no need to worry.

The inescapable make-or-break effort had been made, the risks had been taken, and surmounting a perilous moment caused by outside events, it had succeeded. The annually increasing rate of development continued, buoyantly and confidently now. Having caught up with the numbers and the years, it was in pace with people and time.

In 1976 international dialling was introduced, causing a 22 per cent rise in the number of overseas calls, 2,000,000 outgoing calls in one year. In 1979 there were 1,100,000 exchange lines (the British Post Office's forecast to within a few months) and 1,500,000 actual telephones. In 1983 there were 1,500,000 exchange lines and 2,000,000 telephones.

For a population of 5,233,000, quite a lot of telephones.

The findings of the Commission of 1975 applied, in reality, to conditions which no longer existed when they reported. Certain points stand, however. Of Charles Male they wrote that he 'took over a company affected by the mismanagement of previous years and, by the force of his character, pulled it together and piloted it through an era of extraordinarily rapid expansion. It is probably true to say that compared with the Company under his managerial guidance, no telephone administration in the world has expanded at such a rate — a tremendous achievement.'

Of the public complaints aired, they observed that the telephone had become 'such an essential part of business and social life in Hongkong' that the public expected a consistently excellent service, and complained at the slightest departure from it. A large proportion of the so-called faults, however, were caused by the subscribers themselves — 'mis-dialling, children playing with telephone handsets, subscribers leaving their handsets off the hook to avoid being disturbed, and so on'. Referring to there being no message rate, allowing for 'almost unlimited free use of the telephone', they opined that much of the trouble had arisen from taking a good thing for granted.

To which it must be added that complaint about public utilities — and all of them went through storms at one time or another — provided people with a forum for airing general grievances. No such storms arose when business was booming and grievances were few.

Chapter 26

Two Dimensions

ALREADY in 1976, and throughout the 1980s, the degree of Hongkong people's use of telecommunications, and the related demand for them, were more intense than probably any other city in the world. Demand and degree of use accelerated. For example, it took twenty years (1959–79) for the number of telex subscribers to reach 10,000, yet only four years to achieve 20,000, in 1983. For the cable and telephone companies, when improving a service or introducing innovations, such as data and facsimile services, those concerned had to be prepared for a public response which anywhere else would be considered abnormal.

A point was reached when there was more message-bearing equipment in Hongkong than in any place of comparable size on earth, nearly all of it being used to a prodigious extent.

Hongkong people had no idea that there was anything unusual about the degree to which they used the services until 1981, when an astonishing figure was published. In Britain that year, Cable and Wireless was denationalized and became a public limited company. With shares being offered to the public, more than had been known about the company's operations for many years became general knowledge. It was an enormous company. Despite having lost, by nationalization in 1947, the right to provide services to much of the former British Empire, it had made up for this elsewhere. In 1981 it operated in 29 countries. With its fleet of cable ships, it operated on and beneath all the world's oceans. And 83 per cent of its profits were generated in Hongkong.

At first sight it seemed as if so amazing a fact must have been caused by some strange brand of commercial coercion, whereas in truth it had been generated by determination to keep abreast of seemingly insatiable Hongkong public demand, of which that figure of 83 per cent gave a nice demonstration.

At the time of the denationalization a local company, Cable and Wireless (Hongkong) Ltd., was formed, in which the

Hongkong Government had a 20 per cent ownership. Since 1965, the Government had been pressing for Cable and Wireless to have a seat on the Hongkong Telephone board. Even at that relatively early date the government view, never formally expressed yet understood, was that the colony's best interests would be served if cables and telephones combined, provided the Government had a stake in such a combination.

Cable and Wireless in London had for some time been wanting to buy into the Hongkong Telephone Company, not an easy business in that more than half the shares, at any given time, were held by small investors. In 1983 the Hongkong Land Company, in temporary difficulties, sold their telephone shares to Cable and Wireless. This gave the latter a 34.7 per cent ownership, making them the largest shareholder, yet just short of being in a position to make a general offer.

The following year, 1984, the remaining large shareholder, a member of a prominent Chinese banking family — traced back in time, this was one of the oldest shareholdings — agreed to sell. Cable and Wireless bought, thus acquiring voting control of Hongkong Telephone, which that year became a subsidiary of Cable and Wireless, London.

In 1988 the Hongkong Telephone Company and Cable and Wireless (Hongkong) Ltd. combined to form Hongkong Telecommunications Ltd. The Cable and Wireless Group, London, held 80 per cent of the shares, the Hongkong Government 11 per cent, and the public 9 per cent. In financial terms this was the largest company ever listed on Hongkong's Stock Exchange. Within a matter of months it became the first Hongkong-listed company to be listed on the New York Stock Exchange.

* * *

Fourteen years before this, while Hongkong was in the throes of the boom and crash of 1974, the telephone line to Canton unexpectedly improved. Cable and Wireless had laid a land cable between Hongkong and Canton, a development of more significance than would appear.

Telephone communication between the two cities, severed in 1938, had been re-established in 1945 by providing Canton with a British Army surplus VHF set which Cable and Wireless had managed to get hold of. The public service, opened in January 1948, was reasonably good until the fall of the National

Government in October 1949. Since then, nothing much in the way of improvements had occurred. By 1973, in fact, calling Canton was an experience evoking the days when the ear-piece was in one hand, the mouth-piece in the other.

In 1974 there was a change of policy in China in favour of modernization, which by inference allowed for a degree of foreign investment participation, and thus for improved contacts with the outside world. The Cable and Wireless management in Hongkong, exceptionally well informed, learned of this policy change long before it was publicly known. They opened negotiations with Canton, and laid the much-needed cable. A new phase of telecommunications assistance to China had begun, not dissimilar to that inaugurated by the Danes in the nineteenth century, and to a China which, alike under imperial and communist rule, had been sealed from outside influences and was consequently out of date.

From 1981, the year when China's modernization policy went into full drive, improvements came fast. In that year it became possible to dial Canton from Hongkong, and a mission sent to Canton reached an agreement to introduce a 740-channel microwave communication, a great advance which was in operation by October 1983. Cable and Wireless then provided assistance in establishing microwave transmission between Canton and Swatow, and between Canton and Hainan Island, all three linked to Hongkong and operative in 1986. The Swatow link was then extended to Fukien province.

A high-capacity cable with optical fibre transmission to Kwangtung province was completed in October 1988. In the same year a company was formed to provide a telecom satellite for Asia including China. As the year 1989 came in, it looked as if China was entering a major new era of development.

It furthermore looked as if Hongkong, already recognized as a communications centre for Asia, would be the focal point for this development in South China. This of itself would lend assurance to Hongkong's continuity of character when it ceased to be a British colony in 1997.

On 4 June 1989, in Peking, the suppression of peaceful student demonstrations by tanks and infantry was followed by a clampdown on intellectual activity, and had the effect of throwing into doubt the future of the entire modernization programme, despite official assurances from Peking that modernization would continue.

Seven months later, as this book goes to press, it seemed certain that the telecommunications developments previously envisaged for China would be going ahead as planned. It is greatly to be hoped that they will. For China's integration into the world of the future which is already upon us, they mean a great deal.

Index